Understanding Complex Systems

Founding Editor
Scott Kelso

Series Editors

Henry D. I. Abarbanel, Institute for Nonlinear Science, University of California, San Diego, La Jolla, USA

Dan Braha, New England Complex Systems Institute, University of Massachusetts, Dartmouth, USA

Péter Érdi, Center for Complex Systems Studies, Kalamazoo College, Kalamazoo, USA
 Hungarian Academy of Sciences, Budapest, USA

Karl J. Friston, Institute of Cognitive Neuroscience, University College London, London, UK

Sten Grillner, Department of Neuroscience, Karolinska Institutet, Stockholm, Sweden

Hermann Haken, Center of Synergetics, University of Stuttgart, Stuttgart, Germany

Viktor Jirsa, Centre National de la Recherche Scientifique (CNRS), Université de la Méditerranée, Marseille, France

Janusz Kacprzyk, Systems Research Institute, Polish Academy of Sciences, Warsaw, Poland

Kunihiko Kaneko, Research Center for Complex Systems Biology, The University of Tokyo, Tokyo, Japan

Markus Kirkilionis, Mathematics Institute and Centre for Complex Systems, University of Warwick, Coventry, UK

Jürgen Kurths, Nonlinear Dynamics Group, University of Potsdam, Potsdam, Germany

Ronaldo Menezes, Department of Computer Science, University of Exeter, Devon, USA

Andrzej Nowak, Department of Psychology, Warsaw University, Warszawa, Poland

Hassan Qudrat-Ullah, School of Administrative Studies, York University, Toronto, Canada

Linda Reichl, Center for Complex Quantum Systems, University of Texas, Austin, USA

Peter Schuster, Theoretical Chemistry and Structural Biology, University of Vienna, Vienna, Austria

Frank Schweitzer, System Design, ETH Zürich, Zürich, Switzerland

Didier Sornette, Institute of Risk Analysis, Prediction and Management, Southern University of Science and Technology, Shenzhen, Switzerland

Stefan Thurner, Section for Science of Complex Systems, Medical University of Vienna, Vienna, Austria

Springer Complexity is an interdisciplinary program publishing the best research and academic-level teaching on both fundamental and applied aspects of complex systems—cutting across all traditional disciplines of the natural and life sciences, engineering, economics, medicine, neuroscience, social and computer science.

Complex Systems are systems that comprise many interacting parts with the ability to generate a new quality of macroscopic collective behavior the manifestations of which are the spontaneous formation of distinctive temporal, spatial or functional structures. Models of such systems can be successfully mapped onto quite diverse "real-life" situations like the climate, the coherent emission of light from lasers, chemical reaction-diffusion systems, biological cellular networks, the dynamics of stock markets and of the internet, earthquake statistics and prediction, freeway traffic, the human brain, or the formation of opinions in social systems, to name just some of the popular applications.

Although their scope and methodologies overlap somewhat, one can distinguish the following main concepts and tools: self-organization, nonlinear dynamics, synergetics, turbulence, dynamical systems, catastrophes, instabilities, stochastic processes, chaos, graphs and networks, cellular automata, adaptive systems, genetic algorithms and computational intelligence.

The three major book publication platforms of the Springer Complexity program are the monograph series "Understanding Complex Systems" focusing on the various applications of complexity, the "Springer Series in Synergetics", which is devoted to the quantitative theoretical and methodological foundations, and the "Springer Briefs in Complexity" which are concise and topical working reports, case studies, surveys, essays and lecture notes of relevance to the field. In addition to the books in these two core series, the program also incorporates individual titles ranging from textbooks to major reference works.

Indexed by SCOPUS, INSPEC, zbMATH, SCImago.

Giorgos Tsironis

Artificial Intelligence and Complex Dynamical Systems

Springer

Giorgos Tsironis
Department of Physics
University of Crete
Heraklion, Greece

ISSN 1860-0832   ISSN 1860-0840 (electronic)
Understanding Complex Systems
ISBN 978-3-031-81945-2   ISBN 978-3-031-81946-9 (eBook)
https://doi.org/10.1007/978-3-031-81946-9

© The Editor(s) (if applicable) and The Author(s), under exclusive license to Springer Nature Switzerland AG 2025

This work is subject to copyright. All rights are solely and exclusively licensed by the Publisher, whether the whole or part of the material is concerned, specifically the rights of translation, reprinting, reuse of illustrations, recitation, broadcasting, reproduction on microfilms or in any other physical way, and transmission or information storage and retrieval, electronic adaptation, computer software, or by similar or dissimilar methodology now known or hereafter developed.
The use of general descriptive names, registered names, trademarks, service marks, etc. in this publication does not imply, even in the absence of a specific statement, that such names are exempt from the relevant protective laws and regulations and therefore free for general use.
The publisher, the authors and the editors are safe to assume that the advice and information in this book are believed to be true and accurate at the date of publication. Neither the publisher nor the authors or the editors give a warranty, expressed or implied, with respect to the material contained herein or for any errors or omissions that may have been made. The publisher remains neutral with regard to jurisdictional claims in published maps and institutional affiliations.

This Springer imprint is published by the registered company Springer Nature Switzerland AG
The registered company address is: Gewerbestrasse 11, 6330 Cham, Switzerland

If disposing of this product, please recycle the paper.

*A long due remembrance of Giorgos Tsarouhas
and Stephanos Pnevmatikos*

Preface

Research in *Complex Systems* goes as far as Poincaré and his handling of the three body problem and even before that. The twentieth century became a witness of great advances in nonlinear physics and mathematics especially after the advent and extended use of computers in science. Nonlinear problems are notoriously difficult to tackle analytically and thus the use of computers gave great boost in their understanding. The term "Complexity" usually refers to subjects traditionally know as *nonlinear phenomena* and includes chaos, solitons, fractals, and related topics. The area has become autonomous and addresses many interdisciplinary topics ranging from environment and weather to biology and medicine. Its tools include nonlinear mathematics, probability and statistics, quantum mechanics, etc. but most importantly it relies on computational methods. The recent emergence of *Artificial Intelligence* or *AI* as a potent tool in data science provides a rich set of possibilities for its application in complex systems. *Machine Learning* is part of AI and includes basic rules that enable the handling of data *by example*. This is drastic new approach in science that employs previous knowledge in order to *learn* a computational system to handle new and unknown cases. The expectation is that this type of approach may lead to new knowledge and understanding of complex systems that perhaps will make them more predictable and more manageable.

The aim in the book at hand is to present a number of cases that show exactly that, viz. that AI and machine learning can play a very important role in complex systems and assist in their analysis. The topics chosen fall in the domain of interest of the author while being general problems with a broad scope. They include biological and medical applications, dynamical chaos, discrete systems and coherence, spatiotemporal chaos, and quantum processes. This is a rather broad pool of problems within complex systems that gives a good perspective as to the applicability of AI in physical and biological complexity. One of the main ideas present in this work is the use of analytically solvable nonlinear problems as test-beds of machine learning. If AI can handle well problems that we know how to solve analytically, then it can be applied with confidence in problems we do not know how to solve. This approach gives additional impetus to the blending of AI with complex systems. In order to appreciate this methodology the book contains a succinct introduction to the basic techniques and methods of machine learning. This introductory part contains also basic code written in *Python* that solves the posed problems and gives the methodology used in machine learning. Although this part is far from exhausting the topic, it nevertheless gives most tools necessary for applications in complex systems.

The outline of the book is split in several parts that follows these lines. The first part focuses almost exclusively in basic machine learning methods. In the first chapter we give a short introduction to complex systems and also basic machine learning methodology and general ideas. In Chap. 2 we present two important methods, viz. regression and classification. We use regression generally with data while classification is applied to objects that need to be categorized into classes. In Chap. 3 we expose basic techniques including principal component analysis, singular value decomposition, etc. and also detail the metrics used to evaluate machine learning performance. This ends the first part of the book that gives more general ideas of machine learning. In the second part we focus exclusively on artificial neural networks. In Chap. 4 we introduce neural networks and discuss deep learning. In Chap. 5 we present very powerful types of neural networks such as convolutional and recurrent neural networks and show how they can be applied in relatively simple problems. In Chap. 6 we introduce autoencoders and also discuss reservoir computing. We close this chapter with a very interesting technique termed physics informed machine learning that is very powerful when blending data with modeling equations is necessary. With this topic we close the second part and also the pure machine learning part of the book. The readers who have followed this part and run the codes provided should have a reasonably clear idea on the basics of machine learning that are necessary for the more physical part of the book.

Part III of the book deals with discrete nonlinear systems and in particular with the famous Discrete nonlinear Schrödinger or DNLS equation. This is a basic equation that describes the physics that localized, cubic nonlinearity introduces in condensed matter physics, in photonics as well as other disciplines. Two models of the DNLS equation are exactly solvable and thus play an important role in the use of machine learning in nonlinear systems. The first is the nonlinear dimer system, i.e., a system with two sites that can be solved analytically in terms of elliptic functions. The second is that of the targeted energy transfer model; it is a model for nonlinear resonances that can also be solved analytically in the case of two units. Then in Chap. 7 we introduce the DNLS equation, describe its basic properties, and present the dimer analytical solutions. In Chap. 8 we use a machine learning approach to *learn* these solutions numerically through a machine learning optimization method. We then compare analytical and AI-based dimer solutions and observe how well the latter match the former. Since the method is successful, we apply it in a nonanalytically solvable case of a trimer. Finally, in Chap. 9 we focus on the nonlinear resonance in the targeted energy transfer in a donor-acceptor system and show how well machine learning captures the essence of the resonance. Subsequently we extend this approach to a nonanalytically solvable case in a trimer system. In this part of the book then we employ the basic ideology behind the use of AI in nonlinear systems, viz. once it works with analytically solvable cases we extend it to more complex cases. One may then view this approach as a type of "perturbation" method where we learn the physical system in cases we know well and then explore the cases we cannot handle easily with the standard methods.

In Part IV of the book we focus on systems with larger complexity and variability. In Chap. 10 we use autoencoders in order to study standard chaotic systems such as the Lorenz system and others. We use the information bottleneck of the autoencoders as a layer that can provide minimal information for the embedding properties of the chaotic data. In Chap. 11 we describe chimeras that are spatiotemporally coherent modes which coexist with chaos and show how useful machine learning is in predicting their time evolution. This is a nontrivial feat due to the complexity of the modes and is accomplished when additional information is furnished in the AI methods used. In Chap. 12 we focus on branching phenomena that stem from nonlinearity and stochasticity and describe how they can be analyzed and predicted through AI methods. Finally in Chap. 13 we show how the use of convolutional neural networks can be used in order to classify discrete breathers. These are discrete nonlinear modes in lattice systems that are periodic in time but localized in space. After finishing this part we have a good idea on how machine learning can be utilized for analysis and predictions in chaotic and spatiotemporal models.

In Part V we switch gears and enter the quantum realm with two applications. The first in Chap. 14 deals with the quantum version of the targeted energy transfer model. Here again we exploit analytical results in the simple donor-acceptor system in order to learn an AI system and subsequently use it in cases where no results are available. The ideas thus of testing machine learning with exact models before applying it in cases we cannot solve analytically work in quantum problems as well. The second application is related to quantum computing and in particular to error correction in a simple qubit trimer. Application of quantum autoencoders as detailed in Chap. 15 gives practical results that are utilized in this and larger qubit systems.

In Part VI we move to a completely different domain that is related to complex system applications in biology and medicine. In Chap. 16 we describe the famous Hodgkin-Huxley nonlinear propagation model and show how this is applied to action potential propagation in the heart. We also show how a form of a cardiogram is generated by this pulse before we move in Chap. 17 to apply machine learning in cardiology. We use simple trees and forests and show the type of information one can extract from the cardiogram. We focus on an "extreme" and rather nonintuitive case where hypertension detection can take place from the electrocardiogram information. Finally for this part in Chap. 18 we focus on viral spreading and in particular use COVID-19 data in a machine learning model that used physics informed neural networks. We discuss the success and limitations of this model and describe further extensions. The final Part VII consists of two chapters; the first one, i.e., Chap. 19, brings us up to date with the developments that lead to the avalanche of important awards, such as the Nobel Prize, to contributions in artificial intelligence during the fall of 2024. The book is summarized in the final Chap. 20 where also thoughts about the prospects of further work are presented. In order to make the exposition more complete we describe in the Appendices details on the workings of TensorFlow and Keras as well as mathematical details about elliptic functions and their use.

This exposition grew after a special research course I gave at Harvard University during the spring semester of 2021–2022. This was a very exciting experience that attracted numerous questions and discussions. The following spring semester of 2022–2023 I taught the course in hybrid mode at the University of Crete while a large contingency of students and scientists primarily from Serbia also attended it. This mode was also enjoyable and engaging. Since my department in Crete gave me another teaching assignment the spring semester of the following year I decided to isolate myself in my mountain village in Central Greece and try to put the course "in paper." This is the end result of this effort. There are a number of people I wish to thank who directly or indirectly helped in this project. First and foremost Professor Efthimios Kaxiras of Harvard University for the organization of the course and the strong interaction with him and his group on many topics in physics and machine learning. Special thanks to Professor Edward Ott of Maryland University with whom I had a very interesting long walk back in 2017 and that led to early ideas on the use of machine learning in nonlinear systems. I wish to also acknowledge Professor Pavlos Protopappas' contribution to my knowledge in machine learning both through his very successful course at Harvard that I attended and also through direct discussions. My former student Marios Mattheakis had an important role in the beginning works as did my long term collaborator Georgios Neofotistos. Most work presented in this book was done in direct collaboration with Georgios Barmparis who played an important role in bringing precise machine learning ideas in complex systems. I wish to also thank my student Natalya Almazova as well as my former student Eleni Angelaki for important contributions and ideas in chaotic and biomedical systems, respectively, as well as my long-term collaborator Nikos Lazarides for valuable work in the latter topic. The former undergraduate students Georgios Arapantonis, Jason Andronis, Alkis Chalkiadakis, Myron Theocharakis, and Thomas Dogkas did great work in their undergraduate theses that is in part included here. Special thanks to my editor at Springer Sam Harrison for his promptness and patience.

Cambridge, MA, USA   Giorgos Tsironis
Merkada, Fthiotis, Greece
September, 2024

Contents

Part I Basic Machine Learning Methods

1 Complex Systems and Machine Learning 3
   1.1   Introduction 3
   1.2   Complex Nonlinear Systems 4
         1.2.1   Integrability and Chaos 5
         1.2.2   Solitons and Coherent Structures 8
         1.2.3   Discrete Nonlinear Schrödinger Equation 10
   1.3   Machine Learning Primer 12
         1.3.1   Supervised Learning 14
         1.3.2   Unsupervised Learning 15
         1.3.3   Semisupervised Learning 15
         1.3.4   Reinforcement Learning 16
   1.4   Data Handling and Generalization 16
         1.4.1   Batch Learning 17
         1.4.2   Overfitting and Generalization 17
         1.4.3   Training, Validation, and Testing 18
   1.5   Conclusion 19
   1.6   Summary 20
   References 20

2 Regression and Classification 21
   2.1   Introduction 21
   2.2   Regression 21
         2.2.1   Linear Regression 22
         2.2.2   Nonlinear Regression 25
   2.3   Classification 27
         2.3.1   Logistic Regression 28
         2.3.2   Decision Trees 32
   2.4   Support Vector Machines 33
         2.4.1   Mathematical Formulation 34
         2.4.2   Practical Implementation 35

	2.5	Conclusion	37
	2.6	Summary	38
		References	38

3 Data Manipulation Techniques ... 39
- 3.1 Introduction ... 39
- 3.2 Principal Component Analysis ... 39
  - 3.2.1 Covariance Matrix ... 42
- 3.3 Singular Value Decomposition ... 42
- 3.4 Gradient Descent ... 44
- 3.5 Visualizing Gradient Descent ... 46
- 3.6 Machine Learning Performance Measures ... 47
  - 3.6.1 Classification Metrics ... 48
  - 3.6.2 Regression Metrics ... 50
  - 3.6.3 Other Metrics ... 51
- 3.7 Conclusions ... 51
- 3.8 Summary ... 52
- References ... 52

Part II Artificial Neural Networks and Deep Learning

4 Artificial Neurons and Deep Learning ... 55
- 4.1 Artificial Intelligence ... 55
- 4.2 The OR and XOR Neural Networks ... 56
- 4.3 Deep Learning ... 64
  - 4.3.1 Overfitting and Regularization ... 65
- 4.4 Conclusions ... 66
- 4.5 Summary ... 66
- References ... 67

5 Powerful Neural Network Architectures ... 69
- 5.1 Introduction ... 69
- 5.2 Convolutional Neural Networks ... 70
  - 5.2.1 Convolutional Layer ... 71
  - 5.2.2 Activation Function ... 71
  - 5.2.3 Max Pooling Layer ... 72
  - 5.2.4 Flattening and Softmax ... 72
- 5.3 Recurrent Neural Networks ... 76
- 5.4 Long Short-Term Memory Networks ... 78
- 5.5 Conclusions ... 81
- 5.6 Summary ... 81
- References ... 82

6 Autoencoders and More ... 83
- 6.1 Introduction ... 83
- 6.2 Autoencoders ... 84
  - 6.2.1 Autoencoders and PCA ... 85

Contents   xiii

    6.2.2 Variational Autoencoders... 87
 6.3 Reservoir Computing... 88
 6.4 Physics Informed Machine Learning..................................... 90
 6.5 Conclusion.. 91
 6.6 Summary... 92
 References.. 92

Part III Discrete Nonlinear Models

7 The Discrete Nonlinear Schrödinger Equation............................. 97
 7.1 The DNLS Equation... 97
 7.2 Diabatic Approximation.. 99
 7.3 General Properties of DNLS... 102
 7.4 The Degenerate Nonlinear Dimer.. 104
    7.4.1 Density Matrix Equations....................................... 104
    7.4.2 Localized Initial Conditions.................................... 106
    7.4.3 Weierstrass Solution... 110
    7.4.4 Photonics.. 113
 7.5 The Nondegenerate Nonlinear Dimer.................................... 114
 7.6 Conclusion.. 119
 7.7 Summary... 120
 References.. 120

8 Learning Analytical Solutions... 121
 8.1 Machine Learning for the DNLS Dimer................................ 121
    8.1.1 Localized Initial Conditions.................................... 123
    8.1.2 General Initial Conditions...................................... 125
 8.2 The Nondegenerate Nonlinear Dimer.................................... 126
 8.3 Conclusion.. 128
 8.4 Summary... 128
 References.. 129

9 The Targeted Energy Transfer Model.. 131
 9.1 The DNLS Donor-Acceptor Dimer....................................... 131
    9.1.1 Analytical TET Solution... 132
    9.1.2 Machine Learning with Physics Methodology........ 135
    9.1.3 Results.. 137
 9.2 Transition over a Barrier into a Third State........................... 138
 9.3 Conclusions.. 140
 9.4 Summary... 141
 References.. 142

Part IV Chaos and Spatiotemporal Complexity

10 Dynamical Embedding with Autoencoders.................................. 145
 10.1 Introduction.. 145
 10.2 Description of the Chaotic Systems.................................... 147

		10.2.1	Delay-Coordinate Embedding..	149
	10.3	Autoencoder Methodology..		149
		10.3.1	Data Preparation..	150
		10.3.2	Chaotic Autoencoder Construction..........................	150
		10.3.3	Latent Space Dimension..	152
		10.3.4	Largest Lyapunov Exponent as a Metric	153
		10.3.5	Regularization Parameter Dependent Latent Space Size	154
		10.3.6	Latent Space Dimension as a Function of the Size of the Input Sequence W............................	155
	10.4	Conclusion...		156
	10.5	Summary...		157
	References...			158

11 Chimeras.. 159
 11.1 Introduction... 159
  11.1.1 SQUID Metamaterials with Long Range Coupling......... 160
  11.1.2 Learning and Predicting Chimeras............................. 165
  11.1.3 Observer-Based Neural Network Learning and Prediction... 167
 11.2 Turbulent Chimeras... 168
  11.2.1 Coupled Semiconductor Lasers................................. 169
  11.2.2 Random and Moving Observers................................ 173
 11.3 Conclusion.. 174
 11.4 Summary... 175
 References... 176

12 Branching.. 177
 12.1 Introduction... 177
 12.2 Theoretical Formulation.. 179
 12.3 Numerical Solution of the Hamilton-Jacobi Equation................. 181
 12.4 Machine Learning.. 182
 12.5 Conclusion.. 184
 12.6 Summary... 185
 References... 185

13 Discrete Breathers... 187
 13.1 Introduction... 187
 13.2 Discrete Breathers and Phonons.. 188
  13.2.1 Discrete Breathers.. 189
  13.2.2 Linearized Phonon Modes.. 192
  13.2.3 Creation of Breather and Phonon Modes..................... 193
 13.3 Machine Learning for Breathers and Phonons........................... 193
 13.4 Conclusion.. 195
 13.5 Summary... 196
 References... 196

Part V Quantum Complexity

14 Quantum Targeted Transfer with Machine Learning 199
    14.1 Introduction 199
    14.2 From TET to QTET 200
    14.3 Determination of the Loss Function 201
        14.3.1 Optimization Details 204
    14.4 Machine Learning for QTET 205
        14.4.1 Quantum Dimer 205
        14.4.2 Quantum Trimer 206
    14.5 Conclusion 208
    14.6 Summary 208
    References 209

15 Learning Quantum Systems 211
    15.1 Introduction 211
    15.2 Quantum Neural Network Architecture 212
        15.2.1 Quantum Neural Networks with Conjugate Layers 215
    15.3 Quantum Autoencoders for Bit-Flip Error Correction 216
    15.4 Conclusions 219
    15.5 Summary 220
    References 220

Part VI Biomedical Applications

16 Action Potential Propagation in the Heart 223
    16.1 Introduction 223
    16.2 The Ventricular Fenton-Karma Model 225
    16.3 Numerical Calculation of the Action Potential 228
    16.4 Numerical Evaluation of the Pseudo-Electrocardiogram 230
    16.5 Conclusions 232
    16.6 Summary 233
    References 233

17 Machine Learning Cardiology 235
    17.1 Introduction 235
    17.2 Basics of the Electrocardiogram 236
    17.3 Procedures and Data Handling 238
        17.3.1 Clinical Procedure and Machine Learning 238
        17.3.2 Feature Engineering and Feature Selection 239
        17.3.3 Datasets and Feature Importance 240
    17.4 Results 241
    17.5 Conclusions 243
    17.6 Summary 243
    References 244

18 Epidemiology with Physics Informed Machine Learning ... 245
18.1 Introduction ... 245
  18.1.1 China Data and Early COVID-19 Predictions ... 246
  18.1.2 Time-Gaussian Pandemic Evolution ... 248
18.2 COVID-19 Predictions with Physics Informed Machine Learning ... 250
  18.2.1 Mathematical Manipulations ... 251
  18.2.2 Machine Learning Procedure ... 253
  18.2.3 Short-Term Predictions ... 256
  18.2.4 The Case of Greece ... 257
18.3 Conclusion ... 258
18.4 Summary ... 259
References ... 259

Part VII Conclusion

19 Foundations ... 263
19.1 Introduction ... 263
19.2 The Ising Model ... 264
  19.2.1 Mathematical Foundation of the Ising Model ... 265
19.3 Spin Glasses ... 266
  19.3.1 Ising Spin Glass Systems ... 267
19.4 The Hopfield Model ... 268
  19.4.1 Mathematical Formulation of Hopfield Networks ... 269
19.5 Boltzmann Machines ... 270
  19.5.1 Mathematical Formulation of Boltzmann Machines ... 271
  19.5.2 Learning in Boltzmann Machines ... 272
19.6 The Protein Folding Problem ... 272
  19.6.1 AlphaFold: Solving the Protein Folding Problem with AI ... 273
19.7 Conclusion ... 274
19.8 Summary ... 275
References ... 275

20 Computational Complexity and the Butterfly Effect ... 277
20.1 Closing Remarks and Summary ... 277
20.2 Digital Twins and Nonlinear Physics ... 279
20.3 Universal Approximation Theorem and Chaos ... 279
20.4 Summary ... 279
References ... 280

A Jacobi Elliptic Functions ... 281
A.1 Main Identities ... 282
  A.1.1 Pythagorean-Like Identities ... 282
  A.1.2 Derivative Relations ... 282

		A.1.3	Addition Formulas	282
	A.2		Special Cases	282
B	**Weierstrass Elliptic Function**			283
	B.1	Basic Identities		283
		B.1.1	Derivatives	283
		B.1.2	Differential Equation	284
		B.1.3	Laurent Series	284
		B.1.4	Addition Formula	284
	B.2	Special Cases		284
		B.2.1	Near the Lattice Points	284
		B.2.2	Special Values	284
		B.2.3	Elliptic Curve Relation	285
	B.3	Connection to Other Elliptic Functions		285
C	**Python Programming Language**			287
	C.1	Jupyter Notebooks		288
		C.1.1	Basic Workflow in a Jupyter Notebook	288
	C.2	Advantages of Jupyter Notebooks		288
	C.3	Installing Jupyter Notebooks		289
D	**Introduction to TensorFlow**			291
	D.1	Basic TensorFlow Workflow		291
	D.2	Example: Linear Regression in TensorFlow		292
	D.3	Introduction to Keras		292
	D.4	Basic Keras Workflow		293
		D.4.1	Example: Building a Neural Network with Keras	293
Index				295

Acronyms

AI	Artificial Intelligence
CNN	Convolutional Neural Network
DNLS	Discrete Nonlinear Schrödinger
ECG	ElectroCardioGram
LSTM	Long Short-Term Memory
LVH	Left Ventricular Hypertrophy
MAE	Mean Absolute Error
ML	Machine Learning
PINN	Physics Informed Neural Network
QAE	Quantum AutoEncoder
QTET	Quantum Targeted Energy Transfer
RC	Reservoir Computing
RBM	Restricted Boltzmann Machine
RNN	Recurrent Neural Network
SGD	Stochastic Gradient Descent
SHAP	SHapley Additive exPlanations
SQUID	Superconducting Quantum Interference Device
SVD	Singular Value Decomposition
SVM	Support Vector Machine
TET	Targeted Energy Transfer

Part I
Basic Machine Learning Methods

Chapter 1
Complex Systems and Machine Learning

What Can AI Bring in Complex Systems

Abstract Artificial intelligence is a challenging new discipline of science that tries to combine cognition with computers and computational methodology. A complex system on the other hand is a physical, technological, or biological system that is generally difficult to formulate precisely or solve mathematically and may lead to interesting and sometimes puzzling behaviors. Merging artificial intelligence with complex systems may lead to practical solutions for problems that are both important and generally difficult to handle with other means. We give a small tour of ideas in complex systems and also initiate the study of machine learning. The end result of a machine learning process is a model that has learned to fit data and can determine properties of other data. The model itself is a function or more generally an algorithm that may be applied to datasets. Learning can take place in several ways, the more usual one being supervised, although unsupervised or semisupervised and reinforcement being viable alternatives. Data are the central object in this process, and they are used in order to generate the correct model and also test it. Data is split into a training set and a test set; the former trains the model and determines values of its parameters that are appropriate. The test set part of the data is not used during training, but it is applied to the trained model in order to test how well it works with controlled but unknown data. In order for the model to generalize well, i.e., to be able to work well with new and unknown data, overfitting must be avoided.

1.1 Introduction

The advances in Artificial Intelligence (AI) in the last decades are astonishing, and they seem to have changed completely our way of living. From cell phones to banks, to security, and to social interactions, AI seems to play an increasing role to daily lives of billions of people around the world. There are of course also voices of caution on its general use [1]. On the other hand, it does not seem to have yet penetrated science in a similar profound way. Using machine learning in physics is relatively recent, and certainly it does not appear to have blossomed yet. Why would one want to apply it in physics? There are various levels to the answer of this question. Machine learning may provide methods for handling physical systems that are superior to traditional computational techniques. In machine learning the code does not model the physical system of interest, but it actually *learns* about it. This feature may lead to serious simplifications in the actual representation of the physical system of interest and also the way we can obtain information about the system. Beyond this, one may ask whether the physical system under machine learning control could actually modify itself through the learning process and acquire new features. This Darwinian-like thinking with AI is not unlike the mode of thinking in nonlinear complex systems. In the latter, nonlinearity is a dynamical parameter that allows for drastic system changes and possible adaptation to external influences. Merging AI with complex systems may provide new means of evolution and change for the latter.

1.2 Complex Nonlinear Systems

It is not very easy to define a "complex system" since it depends also on the features we are focusing on. Is a piece of metal a complex system? If we look into the microscopic details of the metal, we find that it comprises a large number of atoms and its physical properties stem from the complicated dynamics of the constituent atoms and electrons. The detailed understanding of these properties involves advanced theoretical and computational methods and many approximations. Yet, if we take this piece of metal and cut it in halves, these two parts will retain essentially all properties of the original piece. Many of the functions of the metal do no depend on this process of halving; for instance, if we use the metal as a wire, cutting it does not change its basic electrical conduction properties.

Consider now a biological macromolecule such as a protein; it is also made of a large—yet not as large as the metal—number of atoms. The protein has many different physical and chemical properties that can be analyzed and understood through exact or approximate methods. These methods may be quite complicated, involved, and cumbersome but in many ways are similar to the methodology we apply to the metal. Yet, there is a significant difference with the metal, viz. the protein has a specific function within the cell. In fact the working of the cell depends on the function of the proteins it includes. If we cut the protein in half, the protein ceases to function, and through it the organization of the cell may become problematic. This specific property, viz. the loss of functionality when drastic changes take place, characterizes complex systems and distinguishes them from the rest. The distinction is neither sharp nor well defined but nevertheless is easy to see when we encounter it. Although biology furnishes the archetypical complex systems, physical systems can also be complex [2, 3]. The weather is one such physical system where long-term understanding and prediction is clearly not yet feasible. It appears that a necessary condition for complexity is the presence of some form of nonlinearity usually in the dynamical equations of motion. Complex systems are generated by nonlinear systems.

1.2 Complex Nonlinear Systems

Nonlinearity is ubiquitous in science, and it would seem at first superfluous to reserve the term for an entire area of study. On the other hand, mathematical techniques that are based on linearity are so powerful and indeed useful that many times tend to overshadow and hide the role of nonlinearity. Linear algebra and linear differential equations are the "bread and butter" of modern science and engineering. At the same time, quantum physics, the powerful intellectual creation of the twentieth century, is "linear." If quantum mechanics that describes all physical processes in the world is linear, what is the reason to invoke "nonlinearity"? One answer is that, linearity at the level of partial differential equations (PDEs), such as the Schrödinger equation, may hide nonlinearity at a different level expressed, for instance, through ordinary differential equations (ODEs). This is particularly true when we consider interacting systems where two or more additional degrees of freedom interact with the variables we are really interested in. In these cases we try to construct a reduced representation where some, perhaps not so relevant, aspects of the problem are eliminated. The apparent linearity of the microscopic world and the Schrödinger equation may then be transformed into nonlinearity at a more intermediate or "mesoscopic" level of description where multiple interactions are taken into account.

1.2.1 Integrability and Chaos

Dynamical systems are described through differential equations such as ordinary differential equations or ODEs or partial differential equations or PDEs, nonlinear maps, and other complex mathematical procedures. In this exposition we focus primarily on differential equations, ODEs, and PDEs. Integrability refers to the possibility of expressing the solutions of differential equations in terms of known functions. For example, in the case of the harmonic oscillator the equation that describes it is a second order ordinary linear differential equation; it is well know that it can be solved through trigonometric functions. Linear systems such as the oscillator present examples of *integrable* systems. In other cases the system is nonintegrable; for instance, the famous Lorenz model [4]:

$$\frac{dx}{dt} = \sigma(y - x) \quad (1.1)$$

$$\frac{dy}{dt} = x(\rho - y) - y \quad (1.2)$$

$$\frac{dz}{dt} = xy - \beta z, \quad (1.3)$$

where x, y, z are dynamic variables that depend on time t and σ, ρ, β are system parameters. Edward Lorenz introduced this system in 1963 as a simplified model for meteorology and showed that the system is *chaotic*. This means that the set of Eqs. (1.1, 1.2,1.3) is nonintegrable, i.e., cannot be solved in terms of known functions. Furthermore, there is *sensitive dependence on initial conditions*, i.e., if we change slightly the initial state, the system wonders around in a completely different part of its state space. A simple Python code for the Lorenz system follows,[1] and the picture of the chaotic attractor that it generates is shown in Fig. 1.1.

```python
import numpy as np
import matplotlib.pyplot as plt
from mpl_toolkits.mplot3d import Axes3D

# Define the Lorenz system
def lorenz_system(state, t, sigma=10.0, rho=28.0, beta=8.0/3.0):
    x, y, z = state
    dx_dt = sigma * (y - x)
    dy_dt = x * (rho - z) - y
    dz_dt = x * y - beta * z
    return [dx_dt, dy_dt, dz_dt]

# Parameters for the simulation
initial_state = [1.0, 1.0, 1.0]   # Initial condition
t = np.linspace(0, 50, 10000)     # Time array for simulation

# Integrate the Lorenz equations
from scipy.integrate import odeint
trajectory = odeint(lorenz_system, initial_state, t)

# Extract the x, y, z coordinates from the trajectory
x = trajectory[:, 0]
y = trajectory[:, 1]
z = trajectory[:, 2]

# Plot the Lorenz attractor in 3D
fig = plt.figure(figsize=(10, 7))
ax = fig.add_subplot(111, projection='3d')
ax.plot(x, y, z, lw=0.5)

# Set plot labels
ax.set_xlabel('X')
ax.set_ylabel('Y')
ax.set_zlabel('Z')
ax.set_title('Lorenz Attractor')

# Show the plot
plt.show()
```

The numerically observed divergence of initially nearby trajectories takes place usually exponentially fast with a rate determined by the largest Lyapunov exponent.

---

[1] All Python codes used in this book can be found at https://github.com/tsironislab.

1.2 Complex Nonlinear Systems

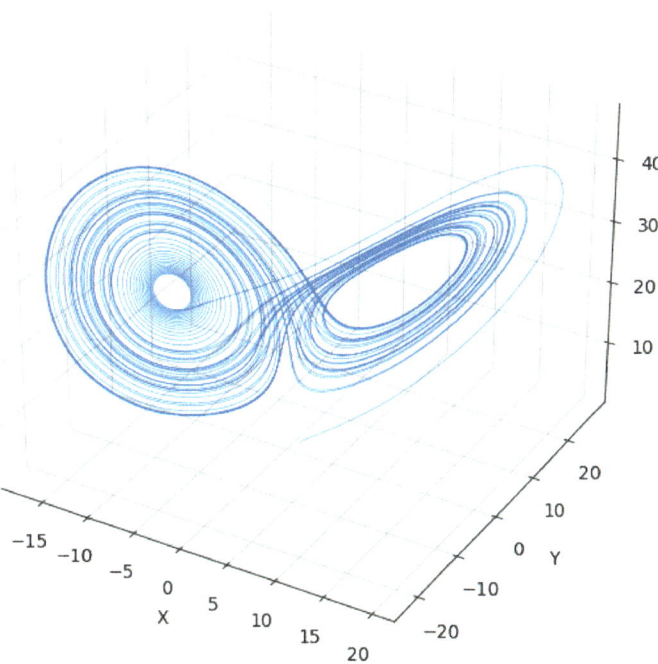

Fig. 1.1 The Lorenz strange attractor is a surface with fractal Hausdorff dimension equal to 2.0627160, i.e., it is slightly larger than 2. A trajectory lying on this surface wanders continuously between the two lobes without a predictable character. In this figure $\rho = 28.0$, $\sigma = 10.0$, and $\beta = 8.0/3.0$ The initial condition used is $(x_0, y_0, z_0) = (1.0, 1.0, 1.0)$

The Lorenz system generates for certain parameter regime a *chaotic attractor*, i.e., a surface with a fractional dimension that has the shape of a butterfly and appropriately is called the *Lorenz attractor*. The shape is related to the sensitive dependence on initial conditions and many times is associated with the famous *Butterfly effect* that is characteristic of complex nonlinear systems. In this effect, the flapping of the wings of a butterfly in Australia can change the weather and its prediction in Boston! Although in this form it might seem as an exaggeration, its true meaning is that in complex nonlinear systems we cannot really predict with a long-term horizon. Assume that we build a powerful computer and we run the model of the world weather focusing on the precise weather in Boston at a given time in the future. The butterfly effect then warns us that if we fail to include in the model seemingly insignificant details, we have a large chance that a small perturbation in Australia might completely ruin the prediction. Chaotic nonlinear systems are very hard to predict; they are in many ways similar to stochastic systems where future evolution can be predicted only in a statistical sense. One interesting question for chaotic dynamical systems is the possibility to use machine learning in order to obtain a better or more useful picture of their future evolution. In fact if machine learning can give a longer and more accurate forecasting horizon in paradigmatic and challenging problems such as weather prediction, then its usefulness will prove to be significant both in theoretical science and in practice.

1.2.2 Solitons and Coherent Structures

In a foundational paper of the 1980s David Campbell posits that the basic aspects of nonlinear science are *fractals, solitons, and chaos* [5]. While chaos is synonymous to sensitive dependence on initial conditions, fractals deal with geometry usually in space. They are self-similar structures that are described through non-integer space dimensions. Intuitively, if the available space is not filled with mass in strict proportion to the available volume, we then have a fractal. A fractal does not have to deal, however, with real mass, and it can be also density of points, as in the case of the Lorenz attractor. In fractals, nonlinearity is of geometric nature while in chaotic systems it is dynamical and typically involves few degrees of freedom. In the other extreme of many dynamical degrees of freedom, we may run into solitary waves and solitons. These are waves that use nonlinearity in order to balance wave dispersion, and as a result a propagating entity is generated that has local, nondispersive nature. Solitons appear in many physical systems such as in water waves and optical and photonic systems, in condensed matter, and in biological systems. They appear as exact solutions of partial differential equations, and they have the property to be essentially transparent one to the other. Solitons are rear but very important solutions of complicated, nonlinear integrable systems.

The Nonlinear Schrödinger or NLS equation is one basic equation appearing both in condensed matter physics and in optics [5]. A standard form of the NLS equation is

$$i\frac{\partial \psi(x,t)}{\partial t} + \frac{1}{2}\frac{\partial^2 \psi(x,t)}{\partial x^2} + |\psi(x,t)|^2 \psi(x,t) = 0, \tag{1.4}$$

where $\psi(x, t)$ is the complex-valued wave function, x is the spatial coordinate, and t is time.

An exact single soliton solution of the NLS equation is

$$\psi(x,t) = A \operatorname{sech}(A(x - vt)) \exp(i(kx - \omega t + \theta)), \tag{1.5}$$

where A is the amplitude of the solitons, v is its velocity, k is the wavenumber, ω is the frequency, and θ is a phase. We notice that the amplitude A enters also in the argument of the hyperbolic function; this means the waves of different amplitudes travel with different velocities, unlike a similar situation in linear waves. The explicit values of the parameters can be found with direct substitution of the solution

1.2 Complex Nonlinear Systems

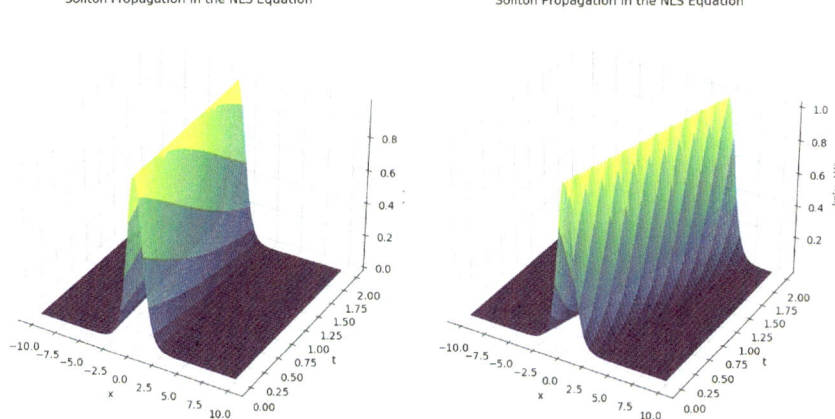

Fig. 1.2 A single NLS soliton moving with velocity 1 (left) and −5 (right)

to Eq. (1.4) [5]. Here is a simple Python code that integrates the NLS equation explicitly with the fourth order Runge-Kutta method and produces moving solitons in Fig. 1.2. We note that the soliton shape is preserved while it moves.

```python
import numpy as np
import matplotlib.pyplot as plt
from mpl_toolkits.mplot3d import Axes3D

# Define the parameters
L = 20.0        # Spatial domain length
N = 256         # Number of spatial points
dx = L / N      # Spatial step size
x = np.linspace(-L/2, L/2, N)  # Spatial grid

T = 2.0         # Total time
dt = 0.001      # Time step
t = np.arange(0, T, dt)  # Time grid

# Parameters for the NLS equation
A = 1.0    # Amplitude of the soliton
v = 1.0    # Velocity of the soliton
k = v / 2  # Wavenumber corresponding to the velocity
theta = 0.0 # Phase

# Initial condition: A moving soliton
psi_0 = A * np.cosh(A * (x - v * 0))**(-1) * np.exp(1j * (k * x + theta))

# Function to calculate the right-hand side of the NLS equation
def rhs(psi, dx):
    psi_xx = np.roll(psi, -1) - 2 * psi + np.roll(psi, 1)
    psi_xx /= dx**2
    return -1j * (0.5 * psi_xx + np.abs(psi)**2 * psi)
```

```python
# Runge-Kutta 4 method
def rk4_step(psi, dt, dx):
    k1 = rhs(psi, dx)
    k2 = rhs(psi + 0.5 * dt * k1, dx)
    k3 = rhs(psi + 0.5 * dt * k2, dx)
    k4 = rhs(psi + dt * k3, dx)
    return psi + (dt / 6) * (k1 + 2 * k2 + 2 * k3 + k4)

# Initialize the solution array
psi = np.zeros((len(t), N), dtype=complex)
psi[0, :] = psi_0

# Time-stepping loop
for i in range(1, len(t)):
    psi[i, :] = rk4_step(psi[i-1, :], dt, dx)

# Plotting the results
fig = plt.figure(figsize=(10, 7))
ax = fig.add_subplot(111, projection='3d')

X, T = np.meshgrid(x, t)
ax.plot_surface(X, T, np.abs(psi)**2, cmap='viridis')

ax.set_xlabel('x')
ax.set_ylabel('t')
ax.set_zlabel('|(x, t)|²')
ax.set_title('Soliton Propagation in the NLS Equation')

plt.show()
```

Solitons are examples of coherent structures that appear in many physical systems and usually described through PDEs. When the equation is not fully integrable as the NLS equation, meaning that there should be a complete set of single and multiple soliton solutions, but we can still find analytically a solution, we then refer to the solutions as solitary waves. We describe a physical phenomenon through continuous nonlinear equations usually, using a continuous approximation. This approach stems from strong coupling between the more elementary units of the medium. On the other hand, if the interaction between the units is weak, the continuous approximation is not so good, and one might have to deal directly with the discrete set of equations describing the dynamics of the units.

1.2.3 Discrete Nonlinear Schrödinger Equation

The Discrete Nonlinear Schrödinger (DNLS) equation is a fundamental nonlinear equation that appears in many areas of physics [6–8]. It is *discrete*, meaning mathematically that it comprises of multiple coupled ordinary differential equations.

1.2 Complex Nonlinear Systems

It is also *nonlinear*, with a specific type of nonlinearity, and thus makes it a significant candidate equation for the study of complex systems with machine learning. It is also related to the *Schrödinger* equation, making it a significant equation for the study of fundamental processes although its scope goes much beyond these. In particular, the DNLS equation is one discrete version of the continuous Nonlinear Schrödinger (NLS) that is one of the main pillars of Nonlinear Science and was introduced previously. If we begin a mathematical treatment of the continuous NLS equation and discretize it, then there are various discrete versions that arise [6, 7]. We will only discuss one specific discretization because it has clear physical interpretation. The specific DNLS equation is important in at least three areas of physics, viz. condensed matter physics, optics and photonics, and Bose-Einstein condensates (BECs). In condensed matter physics it models the propagation of an electron or, more generally an excitation, in a "discrete" medium consisting of atoms or molecules, while the nonlinear term takes approximately into account the exciton-phonon interaction. In optics it describes photon propagation in coupled nonlinear fibers, while in BEC systems the mesoscopic dynamics of the condensate. The form of the DNLS equation we will be using is the following:

$$i\hbar \frac{d\psi_n}{dt} = \epsilon_n \psi_n + V(\psi_{n+1} + \psi_{n-1}) - \chi |\psi_n|^2 \psi_n. \tag{1.6}$$

In the condensed matter interpretation we may think of an excitation that is hopping in a one-dimensional lattice with nearest neighbor interaction term V (Fig. 1.3). The local site energy in each site is ϵ_n; this refers to a single energy state available for the excitation in each molecule where it moves. The basic unknown quantity $\psi_n \equiv \psi_n(t)$ is the probability amplitude to find the excitation at a given lattice site n. If the nonlinear parameter χ was equal to zero for all sites n, then Eq. (1.6) is nothing but the Schrödinger equation in the tight binding approximation. For nonzero χ the DNLS equation becomes nonlinear and acquires entirely new properties that stem from the cubic nonlinear term. In this general form we use here the nonlinear term is assumed to be spatially uniform, i.e., the same in all sites. This is a usual form of the equation; however it is not the most general one. When we discuss the DNLS equation in more detail, we will use χ_n instead of χ; in this case the nonlinearity is local and might change from site to site.

The DNLS equation is fundamental in the understanding of complex nonlinear systems as it encompasses essentially all effects that stem from nonlinearity. For one and two lattice sites is integrable, for few sites is chaotic while for very large, "infinite" sites or degrees of freedom becomes integrable again in the form of the continuous NLS, as we saw previously. In the intermediate range of many degrees of freedom, we find both order and disorder. Order appears in the form of discrete breathers, or intrinsic localized modes, while disorder as spatiotemporal complexity. The study of the DNLS equation is both very interesting and exciting as its results apply in many areas of physics. Implementing machine learning methods to the DNLS equation and other nonlinear systems provides a new frontier for nonlinear

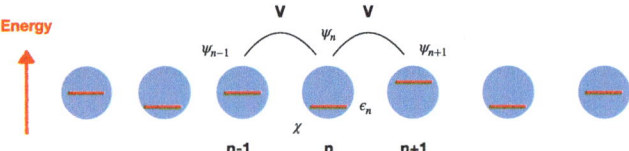

Fig. 1.3 An excitation hopping from site to site with nearest neighbor matrix elements V, local site energies ϵ_n, and nonlinearity parameter χ

science. We hope to learn how far we can go in the understanding of complexity with these knowledge-based methods.

1.3 Machine Learning Primer

The modern scientific method developed during the last few centuries involves experiment and theory. Through the former we obtain data for specific phenomena, while the latter provides mathematical models that explain the data. In recent years the modeling is augmented by numerical methods, while data are also obtained from numerical experiments. In these well-established methodologies modeling has still the role of explaining the data. Machine learning, on the other hand, has a completely different approach and attempts to merge data with models. In this process the data generate the mathematical models directly through well-established procedures. In machine learning the data take the central stage and dictate the underlying modeling, while in a more traditional setting the data are used to fit a model derived independently.

Let us take the example of the weather; this is a truly complex system since it involves a very large number of independent parameters and numerous interconnections among them. We might have at our disposal a large collection of data involving multiple physical quantities of the system such as pressure, temperature, humidity, wind, etc. at various times and in multiple locations. These data are collected through specific, accurate measurements, are then gathered centrally, and can be used essentially in real time. The objective is to predict the weather through the knowledge of the future values of these quantities at various later times. If we have a mathematical weather model at our disposal, we may use these values as *instances* describing the state of the system at a specific space and time frame and evolve it at later times. This process gives a future prediction through the use of available data and the model. In a machine learning setup on the other hand, the data will modify the model itself in order to give the new prediction. The parameters of the model are not given and are continuously modified by the data in order to make the procedure more and more accurate. In this part of the chapter we give introductory information on what is machine learning and how to approach it from a general stand point. We present the basic concepts and tools that lead to successful applications.

1.3 Machine Learning Primer

Fig. 1.4 Standard computations: We write software code in the form of an algorithm, use data, and input both to the hardware of the computer. After debugging the computer produces desired (or undesired) results

In conventional computing central role plays the concept of *algorithm*. The term itself is a "Latinization" of "al-Khwārizmī," i.e., the man from Khiva (modern name), referring to the ninth century astronomer and mathematician Abū Ja'far Muhammad ibn Mūsa al-Khwārizmī, the author of widely translated works of arithmetic and algebra. The algorithm determines a detailed procedure for performing a certain mathematical task, as for instance, calculating the roots of an equation, finding the derivative of a function, etc. In computing, the algorithms used are typically *recurring*, i.e., they are repetitive, and after many executions we may arrive at the desired result. In addition to algorithms, modern computers use also data in the form of integer numbers used as approximations to real numbers. The von Neumann computer architecture requires both algorithms and data in order to obtain results (Fig. 1.4). Both the algorithm and the input data are fed into the computer hardware. The latter then follows a repetitive set of instructions of the software and provides the results.

The essential feature of the procedure is that both input data and the software are completely *fixed*; this means that if for some reason the result is not correct, then we have to throw away the process and start all over again. If the data processing is not acceptable, then we need to rewrite and produce new code, i.e., a new algorithm. If the data represent a reality that the model in the form of the algorithm tries to capture, then the data are not connected to the model and certainly cannot affect it in any direct way. This is where the method based on machine learning differs as seen schematically in Fig. 1.5. The computer hardware is in principle the same; however the software differs drastically from the usual case. Now the algorithm is the end process of the calculation and is generated through the iterative process. While in the input we have the data as in the von Neumann computing, we also include the desired result. The software code is more flexible, it is being exposed systematically to the data, and this interaction leads to an optimized code that is compatible with the data. This approach is completely orthogonal to the usual von Neumann type method and provides as end result the model that connects the data to the expected results.

Machine learning covers a wide range of data applications and involves distinct methods that depend on the type of engagement of a human supervisor, the mode

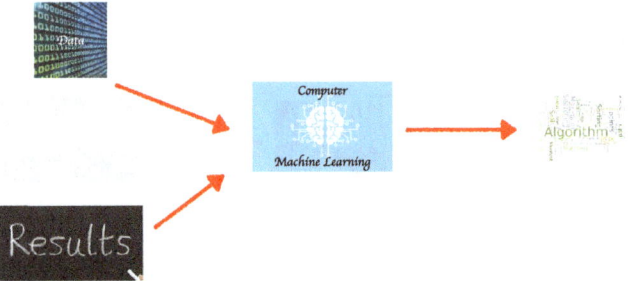

Fig. 1.5 Machine learning computation: The data are connected to results through an imaginative model that is produced through the calculation procedure. The output of this method is the model itself

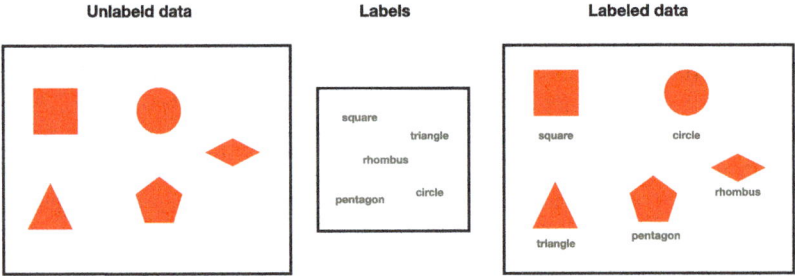

Fig. 1.6 Coupling data with labels forms labeled data. The association of the data to specific labels is done in a supervised way to the training set of the available data

of learning, and the form of pattern detection applied [9]. Supervised learning is a very popular approach, but unsupervised, semisupervised, and reinforcement learning are also distinct alternatives. Learning can occur on the fly or in batch mode, while isolated data instances versus model-based learning present alternative modes. When we have a dataset, we typically split it into the *training set* and the *test set*; we use the former to obtain the model that fits these data and the latter to test whether the generated model works with data not used to produce it. In some cases we also have a *validation set* that is used within the context of the training set.

1.3.1 Supervised Learning

In supervised learning we couple data with desired solutions called *labels*. Every *instance* datum contains then a label which carries the value of the *feature* it represents as in Fig. 1.6. In the process of supervised learning we use labeled data for training, and then we check with the test set how well this data *classification* works. In Fig. 1.7 we show schematically the process of classification.

1.3 Machine Learning Primer

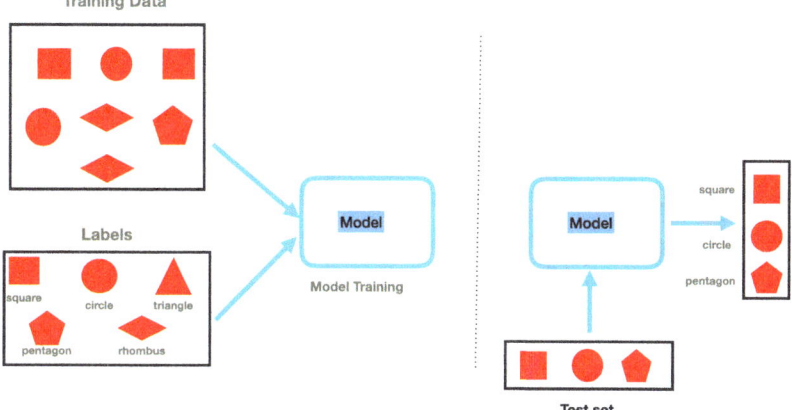

Fig. 1.7 In the process of supervised learning the model is formed through the connection of specific data instances to labels. A square object is associated with the "square" label, etc. Once the training is completed, we feed the model with test set data and expect them to be labeled correctly by the model

1.3.2 Unsupervised Learning

In unsupervised learning the data are not labeled. As they are fed into the machine learning program, they might form groups or clusters that correspond to different aspects. In some sense, these clusters become the labels once the computer learns to separate them. When new data are subsequently presented to the method, they are classified accordingly. In Fig. 1.8 the clustering phenomenon is shown. The unsupervised learning method is very useful in anomaly detection; here a new instance appears that does not fit to the existing patterns. The program then gives a warning about something that is quite different from the training set. This approach is very useful in medicine when we already have a number of data compatible with the "normal" case.

1.3.3 Semisupervised Learning

This is a combination of supervised with unsupervised learning where some aspect of the data has no labels, but the computer can do some labeling. As an example imagine different persons that appear in multiple photos. The model can use unsupervised learning to cluster the different persons recognizing each one in the different photos. Then by labeling one or more of the persons the model can recognize the specific individual in different pictures.

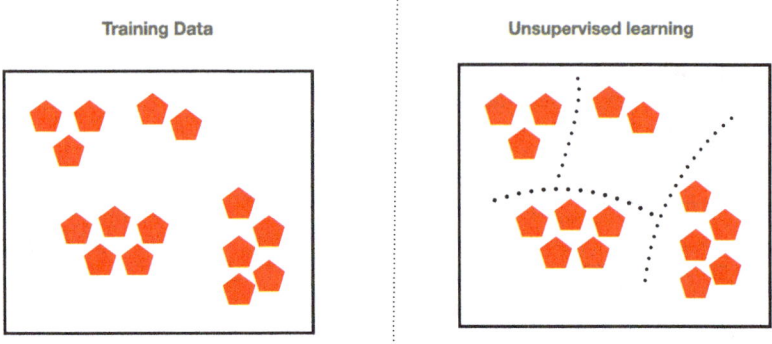

Fig. 1.8 In unsupervised learning we give a set of data and the model recognizes a pattern. In this picture the model performs clustering of data in different groups

1.3.4 Reinforcement Learning

Reinforcement learning is quite different and works similarly to how we train little children, i.e., through rewards and simple forms of punishment. When the child behaves well, we offer ice cream, and when it does not, we give him or her broccoli. In reinforcement learning the *agent*, i.e., the learning system, learns through *rewards* and *penalties* and optimizes this way a strategy termed *policy*. The policy is a function with values that are altered by the rewards and penalties that is eventually minimized (or maximized, depending on the definition) after a reasonable training. The procedure is seen in Fig. 1.9. The agent interacts with the environment and receives positive or negative rewards that help him/her make appropriate decisions after the training phase.

1.4 Data Handling and Generalization

The two central items in machine learning is the data and the specific selection of the model to be used for training and subsequent use. The model can be a simple algorithm such as linear regression or a very complicated function such as a large neural network. In all cases the model has to be trained by the available data and subsequently used on the data to see how well it performs. There are a number of technical issues that permeate all models used.

1.4 Data Handling and Generalization

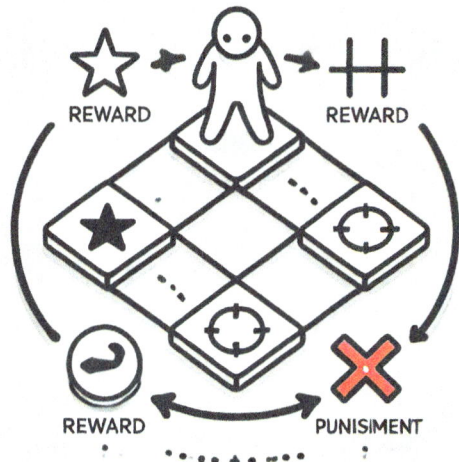

Fig. 1.9 Reinforcement learning: the agent receives training through reward and punishment

1.4.1 Batch Learning

All available data form a *batch* that may be fed in the model to train it. This is a computing intensive process and might not be very useful to use all data at once for the model training. It might be more appropriate to split the data into *mini-batches* and train the model incrementally. This is a more efficient method especially if we need to train using the data in an online mode.

1.4.2 Overfitting and Generalization

When we learn a poem or a piece of text by heart, it is very hard to change words in the text and still recite it. On the other hand, if we have a deeper understanding of a situation and are given new instances of information, we may easily asses if they are related or not with the original setup. *Overfitting* is some form of learning data by heart. Assume we have a set of N data points in the form (x_i, y_i), where the i-th pair may be presented as a data point in a two dimensional x, y space and $i = 1, 2, \ldots, N$; let us take for concreteness N to be 100. We would like to have a model, i.e., a function $y = f(x)$ of one independent variable x that fits the data. It would be easy to construct a polynomial of degree N in such a way that when plotted it passes through every point in the graph. But this would be overfitting, since if we had left out from the dataset few more points and we tried to fit them with the same polynomial, the latter will fail miserably in all probability. Our model is the polynomial function $f(x)$ that has learned precisely the information on the

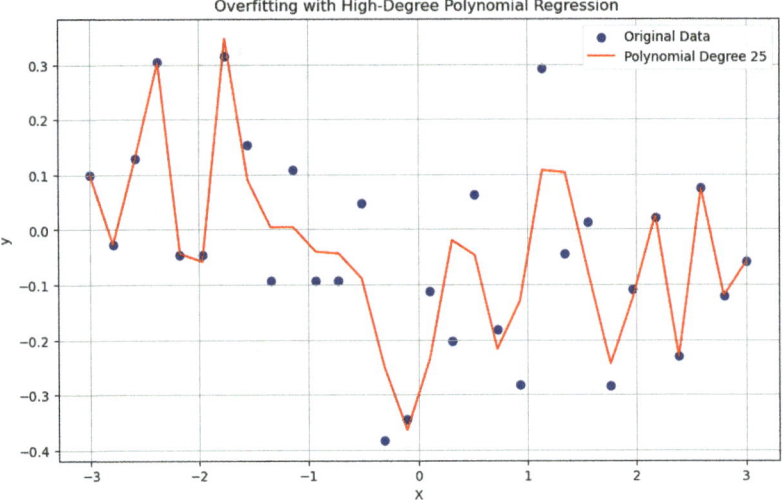

Fig. 1.10 Overfitting a set of random data with a high degree polynomial

first 100 points and simply cannot handle anything more. It would be much more appropriate to fit the data with a lower power polynomial, and this way is able to incorporate more relevant data. In Fig. 1.10 we show fitting and overfitting of data.

In a similar way a model is able to generalize when after the initial training it is able to incorporate and handle data on which it was not exposed before. This is a critical part of a machine learning model, i.e., the ability to handle situations it has not encountered before. Clearly, *generalization* is the main virtue of the human brain and the main reason humans can adopt so well in completely unknown environments and situations.

1.4.3 Training, Validation, and Testing

A machine learning model may have *hyperparameters*, i.e., parameters that need to be selected in order to optimize its performance. This selection can be largely empirical or might also be done in some statistical way. The main objective of a machine learning model is to *learn* an algorithm based on the available data and then use this algorithm successfully with other data. The best way to perform this task is by splitting the data into a *training set* and a *test set*, usually done in an 80%–20% proportion, respectively. We then use the training set to make the model and fix its parameters and the test set that has not been used during training, to test the model. We may use different measures in order to assess the success of the model; one may use the mean square error (MSE) evaluated independently for both the training and the test set. The evaluation is done as a function of the number of epochs, i.e., a measure of internal processing time during which the algorithm does one full sweep of the specific dataset. If the MSE for the test set is much larger than the one of the training sets, then in all likelihood there is overfitting by this model.

The model usually has also hyperparameters, and thus it must be run several times during training for different values of the latter. Assessment of the optimal values can be done empirically or statistically. In the last case, we use one or more *validation sets* that are actually data part of the training set that were not used directly during training. They are used after the training to assess which hyperparameter value is the more appropriate one. Once this value is found, then the final training of the model is done with the full training set that includes the validation set, and then the model performance is tried on the test set.

1.5 Conclusion

Complex systems are nonlinear systems that contain elements of chaos, solitons, and fractals. They provide a framework for understanding natural and biological phenomena through the use of analytical and computational methods. Nonlinear mathematics is notoriously difficult, and the existence of exactly solvable cases is limited. A systematic numerical approach is very useful since it not only gives intuition but also in many times provides the only possible solution to the problem. The use of computational methods is widespread and very efficient. It is hoped that the new computational methods that AI introduces will boost the research in complex systems and provide solutions that are not possible or too difficult to obtain with other means.

Machine learning produces a model that is trained on data and then is tested on data not used during its training. The model has parameters, and these are determined through the learning process. In its simplest form is a function with few unknown parameters that are determined through an automatic optimization process. Care is needed so that the model does not overfit the data. If it does, it cannot really generalize and assess efficiently where new data stand. There are various ways to train the model; the most frequent one is through supervised learning. In this learning we have labels for the data, and we train the model by passing instances of data in the input and the corresponding labels in the output. There is also unsupervised learning where the model makes groupings of the data based on the proximity or not of their features in an appropriate space and semisupervised learning where a mixture of supervised and unsupervised ones is used. In reinforcement learning the model learns by reward or punishment. The data can be handled all at once in a batch or sequentially in mini-batches; the latter is quite efficient also in online learning. The learning itself can be done either by learning instances and then assigning new data based on similarity measures or by directly selecting the appropriate parameters in a model. These two completely different ways of learning are referred to as *instance-based* or *model-based learning*, respectively.

Since the data are the central actor in machine learning, we need to make sure they are of "good quality." This simply means that we need to make sure there are enough of them for the problem at hand and the cases we are interested in well represented. For instance of a binary type choice we need balanced data that contain both cases. In many cases data are missing, and then we need to either remove or augment data. Overfitting is a critical issue in machine learning and needs to be avoided; it is usually easily detected by comparing the errors in the training and test sets. Hyperparameters can help optimize further the machine learning model with optimal values being found through training set validation. Machine learning is a potent quantitative tool for model generation and once mastered can create very useful models that are coupled to data.

1.6 Summary

- Complex systems are dominated by nonlinearity and many interactions. Exact mathematical solutions are rare.
- Nonlinear science includes chaos, solitons, and fractals as basic constituents. Even more complicated features emerge in extended, strongly interacting systems.
- The machine learning model is an algorithm that is trained on part of the data and tested on another part.
- Overfitting is monitored through the comparison of the mean square error of the train and test sets.
- Validation may determine optimal hyperparameter values.

References

1. Max Tegmark, *Life 3.0*, Alfred A. Knopf, New York (2017)
2. G. Nicolis, *Introduction to Nonlinear Science* (Cambridge University Press, Cambridge, 1995)
3. A.C. Scott, *The Nonlinear Universe: Chaos, Emergence, Life* (Springer Science and Business Media, Berlin, 2007)
4. E.N. Lorenz, Deterministic nonperiodic flow. J. Atmosph. Sci. **20**(2), 130–141 (1963)
5. D.K. Campbell, Nonlinear science: from paradigms to practicalities. Los Alamos Sci. **15**, 218 (1987)
6. D. Hennig, G.P. Tsironis, Wave transmission in nonlinear lattices. Phys. Rep. **307**(5–6), 333 (1999)
7. P.G. Kevrekidis, *The Discrete Nonlinear Schrödinger Equation, STMP 232* (Springer, Berlin, 2009)
8. V.M. Kenkre, *Interplay of Quantum Mechanics and Nonlinearity*. Lecture Notes in Physics (Springer, Berlin, 2022)
9. A. Géron, *Hands-on Machine Learning with Scikit-Learn, Keras and TensorFlow*, 3rd edn. (O'Reilly Media, Sebastopol, 2023)

Chapter 2
Regression and Classification

Basic Machine Learning Tools

Abstract Two main tasks of machine learning are regression with specific functions and classification of data into separate classes. Regression is a mathematical method that fits data with a curve, i.e., it passes an optimal curve through a given set of data. In linear regression we use a linear function to provide the best fit to the data. This line does not pass exactly through most of the points as it tries to make an interpolation through them. Nonlinear functions can also be used in regression, but the specific choice depends on the data characteristics and effort to avoid overfitting. Once the fitting function is determined, it can be used in the prediction of new data points. Classification works more with classes that are special groupings of data depending on specific features. Its aim is to learn how to separate the available data into separate classes, and then the model is presented with new data to be able to tell to which class they belong.

2.1 Introduction

Machine learning has a large arsenal of tools and methods that are applied to data. The data can be very simple, such as a simple sequence of numbers, pairs or more generally n-tuples of numbers, vectors, tensors, combinations of numbers with non-numerical data, pictures, videos, and other complex data types. Depending on the question at hand machine learning uses various methods and approaches. In general it uses basic mathematical tools in order to perform tasks that cannot be done simply through other means. Regression and classification are basic methods that are simple yet extremely useful and practical [1–3].

2.2 Regression

Regression is a statistical method used to predict a continuous outcome variable, that is the dependent variable, based on one or more predictor variables that are the independent variables. The goal of regression is to model the relationship between the dependent and independent variables so that one can make accurate predictions on new data.

2.2.1 Linear Regression

A simple, practical method is *linear regression* where we are given a set of pairs of data points (x_i, y_i), $i = 1, 2, \ldots, N$, and we seek the optimal line that passes through them. The line does not necessarily go through the points, in fact for most of the data points it does not. We fit the data with a line because we think that while the data points express some form of a linear functional relationship there are errors involved in the actual measurements. It is thus reasonable this assumed linear relationship to be masked by measurement errors although the latter do not or should not hide the basic functional form connecting the data. If the line has the explicit form $y = h(x) \equiv Wx + b$, where W is the slope and b is the intercept, then the mathematical problem at hand is to find the *optimal* values of W and b that fit the data points. The selection of the best values for the slope and intercept is done through minimization of the mean square error defined as

$$MSE = \frac{1}{N} \sum_{i=1}^{N} (h(x_i) - y_i)^2. \tag{2.1}$$

For the linear regression this minimization problem can be solved analytically, and the two unknown values for the slope and intercept can be obtained through exact formulas. However, it is more instructive and simpler to work directly with the numerical procedure that, additionally, can be generalized easily to cases where there is no analytical solution to the optimization problem. This first and very simple *machine learning* code proceeds as follows: We input the data and assume initial values for the two unknowns of our problem, viz. W and b. We then evaluate the basic quantity that is the mean square error during this first data sweep or *epoch*; the MSE plays the role of the *cost or loss function* in the minimization procedure we follow. The procedure continues iteratively until an optimal pair of (W, b) is found.

```
import numpy as np
import matplotlib.pyplot as plt

X = [0.0, 0.9, 0.21, 0.27, 0.33, 0.42, 0.49, 0.59, 0.72, 1.0]
Y = [0.0, 0.21, 0.61, 0.11, 0.34, 0.30 , 0.21, 0.43, 0.99, 0.71]

mse = []
# Initialization of parameters
N = 10
W = 0.2
b=  0.9
```

2.2 Regression

```
for i in range(1, 1000):
    # Cost function evaluation
    Y_pred = np.multiply(W, X) + b
    error = (Y_pred - Y)**2
    cost = np.sum(error)/N
    # Calculation of changes in W and b
    dW = np.dot((Y_pred -Y ), X)
    db = np.sum((Y_pred - Y))
    mse.append(cost)
    # Updata parameters
    W = W - 0.01 * dW
    b = b - 0.01 * db

    # Repeat for 1000 iterations
    if i%10 == 0:
        print("cost at", i, "iteration = ", cost)

#Plot Cost as a function of # of iterations
print ("W = ", W, "& b = ", b)
plt.plot(mse)
plt.ylabel('MSE')
plt.xlabel("iterations * 10")
plt.savefig('mse.png', format='png')
plt.show
```

After the first sweep the values of *W* and *b* change in a way that in the next epoch the MSE reduces. We iterate until the cost function has a reasonable minimum. The simple code for this procedure that is displayed uses $N = 10$ with a random set of data points. The code produces the mean square error as a function of iteration time or epochs; this is shown in Fig. 2.1. We notice that the mean square error reduces relatively quickly. If you run the code you find that after these 1000 iterations, the code converges in the parameters $W = 0.2240770961239442$

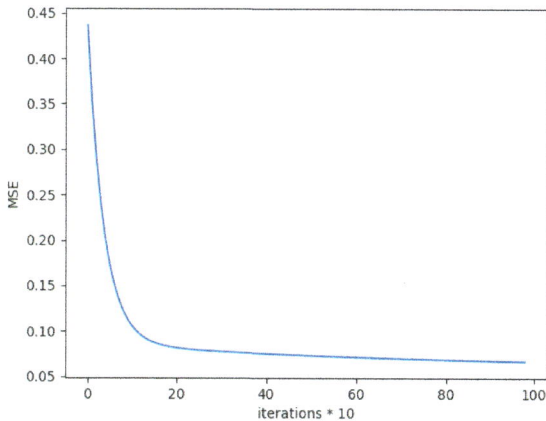

Fig. 2.1 Mean square error evaluated during the linear regression process described in the text as a function of iteration time (in units of ten iteration steps)

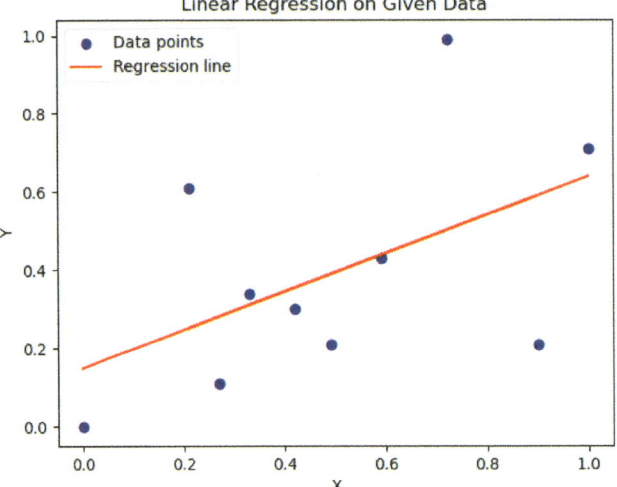

Fig. 2.2 Linear regression for the ten data points shown in the figure

and $b = 0.290524354438205$; these two values determine completely the linear regression process.

The Python code presented above evaluates the mean square error through a simple for loop calculation. While it is instructive to write code that performs intermediate steps in calculations, it is not very economical. This is particularly true in machine learning where detailed code would be long and cumbersome. For this reason we resort to libraries that make the code short and efficient. In the next piece of Python code below we solve the same problem but through the use of the package sklearn that contains functions we need to use in basic machine learning applications. For instructive purposes we use the same data as before and observe the simplicity of this code—it calls the function LinearRegression that performs the task at hand directly. The structure of this code is used repeatedly in machine learning. In Fig. 2.2 we show the outcome of the linear regression calculation. The line that goes through the data points does the best interpolation through the random points generated for this example.

```
import numpy as np
import matplotlib.pyplot as plt
from sklearn.linear_model import LinearRegression

# Given data
X = np.array([0.0, 0.9, 0.21, 0.27, 0.33, 0.42, 0.49, 0.59,
    0.72, 1.0]).reshape(-1, 1)
Y = np.array([ 0.0, 0.21, 0.61, 0.11, 0.34, 0.30 , 0.21, 0.43,
    0.99, 0.71])
```

```
# Perform linear regression
model = LinearRegression()
model.fit(X, Y)

# Predict Y values based on the linear model
Y_pred = model.predict(X)

# Plot the data and the regression line
plt.scatter(X, Y, color='blue', label='Data points')
plt.plot(X, Y_pred, color='red', label='Regression line')
plt.title('Linear Regression on Given Data')
plt.xlabel('X')
plt.ylabel('Y')
plt.legend()

# Display the plot
plt.show()
```

2.2.2 Nonlinear Regression

One main issue in regression is the selection of the model; in the previous example we took by default that a line should fit our data. This is not generally true, and higher polynomials may be used. The selection of the order of the polynomial is crucial and completely user dependent. We thus need to have a good idea about our data before selecting a linear or a higher order polynomial for the regression. If we have N data points, we may fit them exactly with a polynomial of $N-1$ degree, but this is not a good strategy since, as mentioned earlier, it leads to overfitting. In the present example we generate data using a quadratic polynomial while adding also some noise to the data. We thus have the a priori knowledge that a quadratic fit would be the appropriate one. The data generating function is $y = 2x^2 + 3x + 4 + noise$ that in the code enters as Y = 2 * X**2 + 3 * X + 4 + np.random.randn(100, 1) * 5. In the regression we need to recover the three integer coefficients of the quadratic function, viz. $a = 2$, $b = 3$, and $c = 4$. We select a regression with a quadratic function that has these three unknown parameters. In this code we use TensorFlow that is the basic programming tool for machine learning. Some technical details on TensorFlow and Keras and other Python tools for machine learning are described in Appendix A. We note that in the search for the optimal quadratic function that fits the data, TensorFlow performs a gradient search and descents to the values that give the smallest mean square error with the help of the Adam optimizer.[1] The code is shown below while the result in Fig. 2.3. We

---

[1] The Adaptive Moment Estimation is an algorithm for optimization through gradient descent. It is efficient when working with problems involving large datasets or parameters. It is a combination of the gradient descent with momentum algorithm and the RMS propagation algorithm.

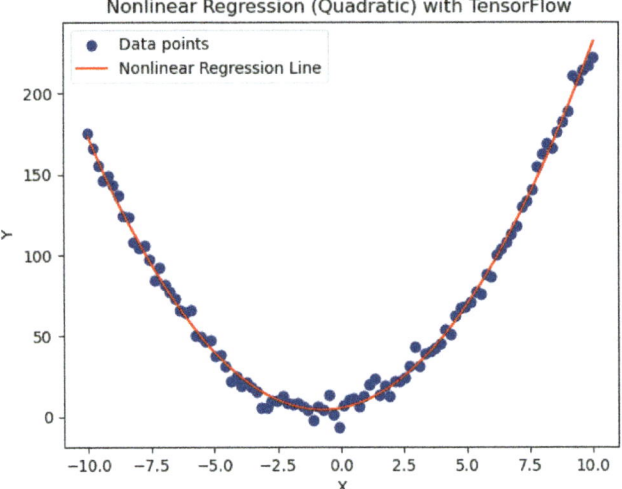

Fig. 2.3 Nonlinear regression for synthetic data with noise

note that the regression is very successful given the fact that a quadratic function was selected. This success provides the general message that we need to have a good idea about the data we are handling before we embark on their analysis. Data intuition plays an important role in machine learning.

```
import numpy as np
import matplotlib.pyplot as plt
import tensorflow as tf

# Generate synthetic data for a nonlinear quadratic function
# y = 2x^2 + 3x + 4 + some noise
X = np.linspace(-10, 10, 100).reshape(-1, 1)
Y = 2 * X**2 + 3 * X + 4 + np.random.randn(100, 1) * 5  # Adding some
 # noise

# Define variables-coefficients for the quadratic model
a = tf.Variable(0.0, dtype=tf.float32)
b = tf.Variable(0.0, dtype=tf.float32)
c = tf.Variable(0.0, dtype=tf.float32)

# Define the nonlinear model: y = a * x^2 + b * x + c
def nonlinear_model(x):
    return a * x**2 + b * x + c

# Define the loss function-Mean Squared Error
def loss_fn(y_true, y_pred):
    return tf.reduce_mean(tf.square(y_true - y_pred))

# Define the optimizer-Adam optimizer
```

```python
optimizer = tf.optimizers.Adam(learning_rate=0.1)

# Training the model
def train_step(X, Y):
    with tf.GradientTape() as tape:
        predictions = nonlinear_model(X)
        loss = loss_fn(Y, predictions)
    gradients = tape.gradient(loss, [a, b, c])
    optimizer.apply_gradients(zip(gradients, [a, b, c]))
    return loss

# Training loop
epochs = 2000
for epoch in range(epochs):
    current_loss = train_step(X, Y)
    if epoch % 100 == 0:
        print(f"Epoch {epoch}, Loss: {current_loss.numpy()}")

# Make predictions using the trained nonlinear model
Y_pred = nonlinear_model(X).numpy()

# Plot the data and the regression line
plt.scatter(X, Y, color='blue', label='Data points')
plt.plot(X, Y_pred, color='red', label='Nonlinear Regression Line')
plt.title('Nonlinear Regression (Quadratic) with TensorFlow')
plt.xlabel('X')
plt.ylabel('Y')
plt.legend()
plt.show()
```

2.3 Classification

Assume that we have datasets that belong to different families of objects; for instance, in a collection of animals we may have cats, dogs, elephants, etc. A class is a given category of objects that are grouped together based on common characteristics. Classes are also referred to as targets or labels. For each element of a dataset at hand, we may know its class membership or, equivalently, its label. Classification is the machine learning process whereby we assign a class or a label to new data that are not part of the original dataset. This is another supervised learning method where the model learns from the training set data to make categorical predictions. The classification can be binary or multi-class and may involve different methods such as linear classifiers, logistic regression, decision trees, SVMs,[2] Bayesian classifiers,[3] kNNs,[4] and deep learning methods. In order to apply these methods we will use the *Iris dataset* that is a dataset with real data used with various classifiers and well tested. It consists of three flower varieties, *Iris setosa*, *Iris virginica*, and *Iris versicolor*, that differ and are classified according to four numbers, i.e., the petal length and width and the sepal length and width. The database is balanced and contains 150 iris flowers. The logistic regression classifier is a probabilistic extension of regression methods and is quite versatile. We will apply this classification method with the Iris dataset.

2.3.1 Logistic Regression

The logistic regression method uses a continuous function to turn linear regression into a categorical separation method. It works with binary or multi-class data and aims at separating them accordingly through the use of the logistic function

$$\sigma(z) = \frac{1}{1 + \exp\{-z\}}. \tag{2.2}$$

The function $\sigma(z)$ plotted in Fig. 2.4 takes values in the range 0, 1 for $z \in (-\infty, +\infty)$ while at $z = 0$ takes the value $\sigma(0) = 1/2$. It is then an appropriate function to turn a continuous variable z into a probabilistic form. It is also called a *sigmoid function* due to its sigma-like shape. The variable z can be determined from a linear regression of the form $z = WX + b$, where W is in general a vector of weights, b the bias, and X the set of features under classification. The outcome of the linear regression is passed through the logistic function $\sigma(z)$ and turned into probability. If we set a certain threshold, for instance, $1/2$, we readily have a probabilistic classifier that separates one class with $\sigma(z) > 0.5$ from a second class with $\sigma(z) \leq 0.5$.

Since the logistic regression is fundamentally a probabilistic classification method, we use a cost or loss function that is reminiscent of entropy. Assume that the estimated probability is given by $p = h_\theta(z) = \sigma(WX + b)$, where θ denotes all parameters in the model and p is the predicted probability. The cost function used is

$$J(\theta) = -\frac{1}{N} \sum_{i=1}^{N} [y_i ln(p_i) + (1 - y_i) ln(1 - p_i)], \tag{2.3}$$

---

[2] Support Vector Machine (SVM) is a powerful supervised learning algorithm used for both classification and regression tasks. It will be analyzed later in the chapter.

[3] A family of probabilistic classifiers based on Bayes' theorem.

[4] k-Nearest Neighbors (kNNs) is a simple, nonparametric, instance-based learning algorithm used for both classification and regression tasks. The prediction for a new data point is based on the "k" most similar training examples that are neighbors in the feature space.

2.3 Classification

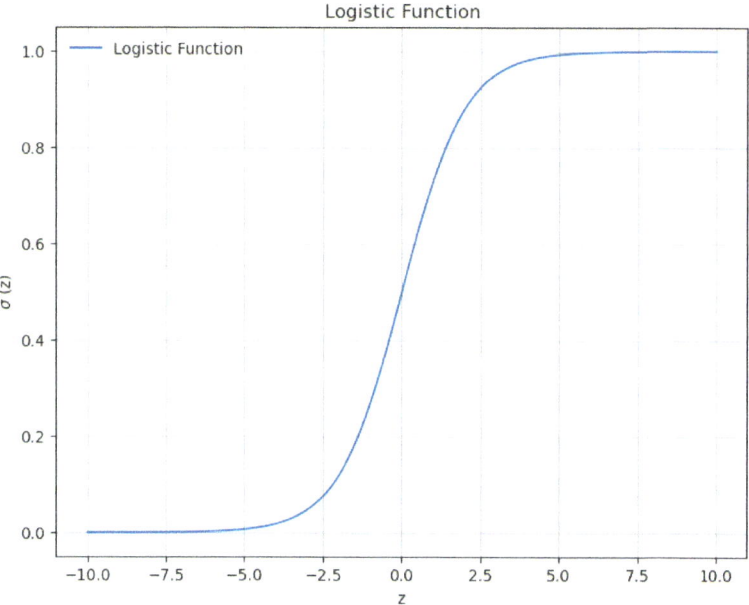

Fig. 2.4 Logistic or sigmoid function $\sigma(z)$ as a function of z

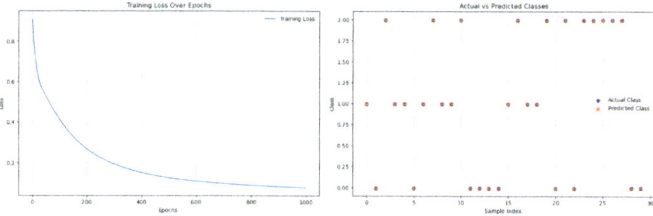

Fig. 2.5 Logistic regression: cost function optimization and iris flower classification

where N is the total number of training examples, θ are the weights of the logistic regression model that we want to learn, y_i is the label of the i-th training example that takes the values 0 or 1, and p_i is the predicted probability that the i-th example belongs to class 1 given by $p_i = \sigma(\theta^T x_i)$, where x_i is the feature vector of the i-th example.[5]

The code that performs the logistic classification for the Iris dataset follows, while in Fig. 2.5 we show the obtained results. We observe a decay of the cost function with the number of epochs and a prefect classification of the flowers in three classes, Setosa, Virginica, and Versicolor.

---

[5] $\theta^T = (W_1, W_2, \cdots W_N, b)$

```python
import numpy as np
import tensorflow as tf
from sklearn.datasets import load_iris
from sklearn.model_selection import train_test_split
from sklearn.preprocessing import StandardScaler
import matplotlib.pyplot as plt

# Load the Iris dataset
iris = load_iris()
X = iris.data  # Features: sepal length, sepal width, petal length, petal
#    width
y = iris.target  # Labels: 0 - Setosa, 1 - Versicolor, 2 - Virginica

# Split the data into training and testing sets (80% training, 20%
#    testing)
X_train, X_test, y_train, y_test = train_test_split(X, y, test_size=0.2,
    random_state=42)

# Standardize the data (mean 0, variance 1)
scaler = StandardScaler()
X_train = scaler.fit_transform(X_train).astype(np.float32)  # Convert to
    float32
X_test = scaler.transform(X_test).astype(np.float32)  # Convert to float32

# Define logistic regression model using TensorFlow (no neural networks)
class LogisticRegressionModel(tf.Module):
    def __init__(self):
        # Initialize weights and biases
        self.W = tf.Variable(tf.random.normal([4, 3], dtype=tf.float32))
        # 4 features, 3 classes
        self.b = tf.Variable(tf.zeros([3], dtype=tf.float32))  # 3 classes

    def __call__(self, X):
        # Linear model: y = XW + b
        return tf.matmul(X, self.W) + self.b

# Initialize the model
model = LogisticRegressionModel()

# Loss function (cross-entropy with integer labels)
def loss_fn(y_true, y_pred):
    return tf.reduce_mean(tf.nn.sparse_softmax_cross_entropy_with_logits
        (logits=y_pred, labels=y_true))

# Define the optimizer (Adam)
optimizer = tf.optimizers.Adam(learning_rate=0.01)

# Training the logistic regression model
def train_step(X, y):
    with tf.GradientTape() as tape:
        # Compute predictions and loss
        predictions = model(X)
```

2.3 Classification

```python
        loss = loss_fn(y, predictions)
    # Compute and apply gradients
    gradients = tape.gradient(loss, [model.W, model.b])
    optimizer.apply_gradients(zip(gradients, [model.W, model.b]))
    return loss

# Training loop with loss tracking
epochs = 1000
loss_history = []  # To store loss values over epochs
for epoch in range(epochs):
    current_loss = train_step(X_train, y_train)
    loss_history.append(current_loss.numpy())  # Track the loss
    if epoch % 100 == 0:
        print(f"Epoch {epoch}, Loss: {current_loss.numpy()}")

# Make predictions (softmax for probabilities)
def predict(X):
    logits = model(X)
    return tf.nn.softmax(logits)

# Evaluate the model on the test set
predictions = predict(X_test)
predicted_classes = np.argmax(predictions, axis=1)

# Calculate accuracy
accuracy = np.mean(predicted_classes == y_test)
print(f"Test accuracy: {accuracy * 100:.2f}%")

# Plot the training loss over epochs
plt.figure(figsize=(10, 6))
plt.plot(loss_history, label='Training Loss')
plt.title('Training Loss Over Epochs')
plt.xlabel('Epochs')
plt.ylabel('Loss')
plt.legend()
plt.grid(True)
plt.show()

# Plot actual vs predicted classes
plt.figure(figsize=(10, 6))
plt.scatter(range(len(y_test)), y_test, label='Actual Class', marker='o',
    color='blue')
plt.scatter(range(len(predicted_classes)), predicted_classes,
    label='Predicted Class', marker='x', color='red')
plt.title('Actual vs Predicted Classes')
plt.xlabel('Sample Index')
plt.ylabel('Class')
plt.legend()
plt.grid(True)
plt.show()
```

2.3.2 Decision Trees

The use of decision trees provides a simple and intuitive classification method involving a tree-like structure; the method can also be used for regression. The data are split into subsets based on feature values, and the process that follows is done in a recursive way. The process starts with the complete dataset that comprises the *root node*. Using the most significant feature available, the root node is split into two or more *branches* that lead into *leaf nodes*. The leaves represent class labels in classification or a continuous variable in the case of regression. The splitting process depends on various algorithms based on functions such as Gini impurity, entropy, etc. The goal is to reach leaves with high degree of purity, viz. the subsets contain mostly data from the given class.

The code that follows shows a decision tree for the Iris dataset.

```python
from sklearn.datasets import load_iris
from sklearn.tree import DecisionTreeClassifier
from sklearn.tree import plot_tree
import matplotlib.pyplot as plt

# Load the Iris dataset
iris = load_iris()
X = iris.data
y = iris.target

# Create a decision tree classifier with a maximum depth of 3 to limit the
#   number of branches
clf = DecisionTreeClassifier(max_depth=3, criterion='gini')
clf.fit(X, y)

# Plot the decision tree
plt.figure(figsize=(12, 8))
plot_tree(clf, feature_names=iris.feature_names,
    class_names=iris.target_names, filled=True,
    impurity=True, rounded=True)
plt.title('Decision Tree with Gini Index')
plt.show()
```

The result of the classification is shown in Fig. 2.6. It is based on the Gini metric. The Gini impurity function measures the frequency of incorrect classifications of randomly chosen elements:

$$Gini = 1 - \sum_{i=1}^{n} p_i^2, \tag{2.4}$$

where p_i is the probability of an element being classified into class i in a total of n classes.

2.4 Support Vector Machines

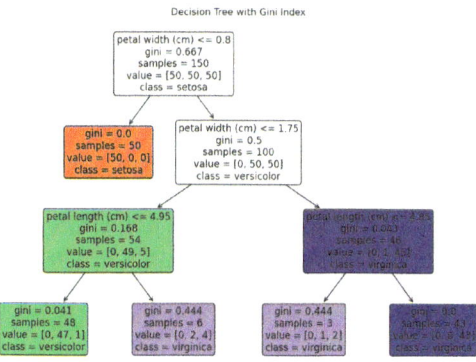

Fig. 2.6 Decision tree classification in the Iris dataset based on the Gini impurity function. The aim is to reach pure states if possible that contain only elements from the same class

Decision trees are very intuitive and simple to use with discrete data but are prone to overfitting. They are the basic elements in ensemble methods such as random forests where the additional statistical inference from the use of multiple trees combines their predictions and gives better results.

2.4 Support Vector Machines

Support vector machine or SVM is a supervised machine learning method that separates data into classes by finding the best boundaries, or *hyperplanes*, that divide them. For two dimensional data, the boundary is a line, in three dimensions a plane while in higher dimensions a hyperplane. The criterion for the best hyperplane is that it leaves the largest margin between the classes, i.e., the classes have the widest gap between them. Pictorially in two dimensions this is seen as a wide highway that separates different types of data lying on each of the two sides. *Support vectors* are the data closest to this hyperplane and are critical in defining the largest gap between the classes that are separated, while the *margin* is the closest distance from the hyperplane to the support vector. The SVM method tries to maximize this margin in order to ensure the largest confidence in the separation between the classes.

When data are linearly separable a linear SVM is used that determines a line (2D) or a plane (3D) that divides the classes and provides the largest margin. If linear separation is not possible, a nonlinear transformation called the *kernel trick* is used that maps the data in a higher dimensional space where they are actually linearly separable. A polynomial kernel may be used, and also a sigmoid or a radial basis function kernel is able to perform the effective linearization of the data.

2.4.1 Mathematical Formulation

It is instructive to give a brief exposition of the linear SVM for binary classification. Assume $X = \{x_1, x_2, \ldots, x_n\}$ represent training data in a d-dimensional space, i.e., $x_i \in \mathbb{R}^d$ is a d-dimensional vector, and $y = \{y_1, y_2, \ldots, y_n\}$ represent corresponding labels where $y_i \in \{-1, 1\}$. The labels in SVM typically take these two values, i.e., ± 1. The goal of linear SVM is to find the optimal hyperplane that separates the two classes, i.e., $y_i = +1$ and $y_i = -1$.

We express a hyperplane in $\mathbb{R}^d$ as

$$w \cdot x + b = 0, \tag{2.5}$$

where $w \in \mathbb{R}^d$ is the *weight vector* that is normal to the hyperplane, $b \in \mathbb{R}$ the *bias* term, and $x \in \mathbb{R}^d$ the vector of the input data. In order for the classification to take place we need a decision rule that places the labels on opposite sides of the hyperplane, i.e.,

$$\hat{y} = \begin{cases} +1, & \text{if } w \cdot x + b \geq 0 \\ -1, & \text{if } w \cdot x + b < 0, \end{cases} \tag{2.6}$$

where $\hat{y}$ is the predicted label for the data.

For a given hyperplane $w \cdot x + b = 0$, the distance of a datum x_i to the hyperplane is given by

$$d = \frac{|w \cdot x_i + b|}{||w||}, \tag{2.7}$$

where $||w||$ is the norm of the weight vector. For margin minimization the data points closest to the boundary must comply with the following constraints: For each point x_i in the $y_i = +1$ class $(w \cdot x_i + b) \geq 1$, while for those in the $y_i = -1$ class $(w \cdot x_i + b) \leq -1$; since the label takes only ± 1 values, these constraints can be combined into one:

$$y_i(w \cdot x_i + b) \geq 1 \quad \forall i. \tag{2.8}$$

Equation (2.8) ensures correct classification of each data point on the opposite sides of the hyperplane and outside the margin boundary. The optimization problem is to minimize $||w||$ or, equivalently, $\frac{1}{2}||w||^2$:

$$\min_{w,b} \frac{1}{2}||w||^2 \tag{2.9}$$

subject to the constraint of Eq. (2.8). This optimization procedure works well when the data are separable and the margin rigid. If this is not the case, then one may

2.4 Support Vector Machines

introduce a *soft margin* that permits misclassification. This is accomplished through the introduction of slack variables $\xi_i \geq 0$ that give some extra freedom in the constraint, viz.

$$y_i(w \cdot x_i + b) \geq 1 - \xi_i \quad \forall i$$

The objective now becomes

$$\min_{w,b,\xi} \frac{1}{2}||w||^2 + C\sum_{i=1}^{n} \xi_i$$

subject to

$$y_i(w \cdot x_i + b) \geq 1 - \xi_i, \quad \xi_i \geq 0 \quad \forall i,$$

where C is a regularization parameter controlling the trade-off between margin maximization and classification error.

2.4.2 Practical Implementation

We now use soft margin SVM in order to classify the iris flower data based on two of their features, sepal length and sepal width. The Python code follows:

```python
import numpy as np
import matplotlib.pyplot as plt
from sklearn import datasets
from sklearn.model_selection import train_test_split
from sklearn.preprocessing import StandardScaler
from sklearn.svm import SVC

# Load the Iris dataset
iris = datasets.load_iris()
X = iris.data[:, :2]  # Take only the first two features (sepal length and
#   sepal width) for easy visualization
y = iris.target

# Keep only the first two classes for binary classification
X = X[y != 2]
y = y[y != 2]

# Split the data into training and test sets
X_train, X_test, y_train, y_test = train_test_split(X, y, test_size=0.3,
    random_state=42)

# Standardize the features (mean 0, variance 1)
```

```python
scaler = StandardScaler()
X_train = scaler.fit_transform(X_train)
X_test = scaler.transform(X_test)

# Create a soft margin SVM (using RBF kernel for flexibility)
model = SVC(kernel='linear', C=1.0)  # Linear kernel, C controls the
#   margin softness

# Train the SVM model
model.fit(X_train, y_train)

# Plotting decision boundaries
def plot_decision_boundary(X, y, model):
    # Define limits for the plot
    x_min, x_max = X[:, 0].min() - 1, X[:, 0].max() + 1
    y_min, y_max = X[:, 1].min() - 1, X[:, 1].max() + 1
    xx, yy = np.meshgrid(np.linspace(x_min, x_max, 500),
        np.linspace(y_min, y_max, 500))

    # Predict class labels for all points on the grid
    Z = model.predict(np.c_[xx.ravel(), yy.ravel()])
    Z = Z.reshape(xx.shape)

    # Plot decision boundary and margins
    plt.contourf(xx, yy, Z, alpha=0.3, cmap=plt.cm.coolwarm)
    plt.scatter(X[:, 0], X[:, 1], c=y, s=50, edgecolors='k',
        cmap=plt.cm.coolwarm)

    # Plot support vectors (these define the margins)
    plt.scatter(model.support_vectors_[:, 0], model.support_vectors_[:,
    #   1],
        s=100, facecolors='none', edgecolors='k', label="Support Vectors")

    plt.title('Soft Margin SVM on Iris Dataset')
    plt.xlabel('Sepal Length (standardized)')
    plt.ylabel('Sepal Width (standardized)')
    plt.legend()
    plt.show()
```

The code loads the Iris dataset that consists of three classes of iris plants (Setosa, Versicolor, and Virginica). Subsequently it selects only the first two flower features X = iris.data[:, :2], viz. sepal length and sepal width, while the target labels y are the species of the iris plants. The command y != 2 filters out the third class (Virginica) from the dataset, leaving only two classes (Setosa and Versicolor). The dataset is then split into training 70% and 30% test sets and after standardizing the features by removing mean and dividing by the standard deviation creates a Support Vector Classifier (SVC) and trains with fit the data. Figure 2.7 shows the decision boundary together with the margin and support vectors for the model separation in two classes.

2.5 Conclusion

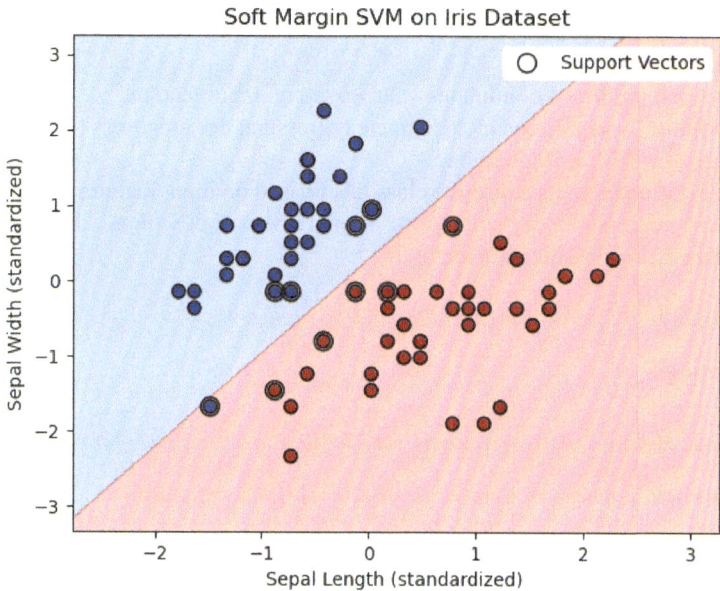

Fig. 2.7 Iris classification results into Setosa and Versicolor based on sepal length and sepal width with a soft margin SVM. The decision boundary separates almost all points in the two distinct classes

2.5 Conclusion

Regression and classification are two basic tasks of machine learning that make predictions but with different types of results. Regression predicts a continuous value through a functional expression that is based on input features. One example is to predict the prices of used automobiles based on their age, manufacturer, location where they were used, etc. In addition to linear regression, decision trees or SVM can also be used for regression. The method gives in the output a numerical value that may be price, time, temperature, pressure, etc.

Classification on the other hand is used to predict discrete class labels based on the input features given in a specific problem. Examples are the classification of emails in two classes, spam or not spam, the prediction of the health status of an individual based on his/her medical data. The most usual algorithms for classification are the logistic regression method, decision trees, SVM, kNN, etc. In the output of a classification method we obtain a category or label such as "spam" or "not spam," "healthy" or "not-healthy," etc.

While regression predicts numeric values, classification assigns labels to data. Both methods use machine learning algorithms in order to learn patterns from the set of data used for training and subsequently generalize in order to make predictions for new and unseen previous data.

2.6 Summary

- Regression predicts a continuous value based on input features.
- Algorithms for regression include linear regression, decision trees for regression, and Support Vector Regression.
- Classification predicts a discrete class label based on input features.
- Logistic regression, decision trees, Support Vector Machines, and k-Nearest Neighbors are classification methods.

References

1. A. Géron, *Hands-on Machine Learning with Scikit-Learn, Keras and TensorFlow*, 3rd edn. (O'Reilly Media, Inc., 2023)
2. I. Goodfellow, Y. Bengio, A. Courville, *Deep Learning* (MIT Press, 2016)
3. O. Campesato, *Artificial Intelligence, Machine Learning and Deep Learning* (Mercury Learning and Information, Dulles, 2020)

Chapter 3
Data Manipulation Techniques

Focus on the Important Aspects of a Data Set

Abstract High dimensional data do not occupy uniformly the space leading to problems in learning models. Using the variance or spread of the data in different high dimensional space directions as criterion one obtains characteristic data directions containing most relevant information of specific features. The principal components determine these directions and simplify the further analysis of data. Use of the singular value decomposition method for rectangular matrices of data with labels assists also in this problem. In order to test the efficiency of methods we use metrics such as the confusion matrix, mean square error, etc that determine how well the trained models work with test data.

3.1 Introduction

Handling of large quantities of data is very challenging especially when data are high dimensional. A number of methods are available that help make the analysis more efficient and precise. Additionally, once the training phase of the model is completed, we always need to know how well it performs on data not used for training. There are a number of metrics that do exactly that, i.e. give information on how the test set of data behaves under the action of the model. In this chapter we study a number of data analysis methods that are very useful in simplifying the process. Furthermore, metrics both for classification and regression are discussed that give information on the efficiency of the models used. Since machine learning is a sophisticated optimization method, we also discuss briefly the process of gradient descent that is routinely used in finding the optimal values for the parameters in the models.

3.2 Principal Component Analysis

Many times we have high dimensional data that need to be placed in spaces of high dimensions in order to be analysed. In these cases we might run into some geometrical trouble if we don't have a lot of data. Let us consider a unit hypercube in n dimensions, where $n = 1, 2, 3, \ldots$ and place data in it. The largest distance in the interior of the hypercube is its diagonal d_n. In one dimension the length of the segment coincides to the diagonal, viz. $d_1 = 1$. In two dimensions the shape is that of a square of unit side and thus the diagonal is simply $d_2 = \sqrt{1^1 + 1^2} = \sqrt{2}$. Similarly in three dimensions the diagonal of the unit cube is $d_3 = \sqrt{1^1 + 1^2 + 1^2} = \sqrt{3}$, etc while in the general case of n dimensions the hyperdiagonal is simply $d_n = \sqrt{n}$. We notice that although the side of the cube is always the same, viz. of length one, as the dimension of the space n increases the diagonal of the cube scales with the square root of this dimension. For $n = 100$ the largest length in the cube is equal to 10, for $n = 10^4$ becomes 100 while for $n = 10^{12}$ is already one million! If we check similarly the total surface s_n that surrounds the cube, we find $s_1 = 0$ since it consists of only two points in one dimension, $s_2 = 4 = 2 \times 2$, $s_3 = 6 = 3 \times 2$, etc. In the general case of n dimension the surface area is $s_n = n \times 2 = 2n$. As the unit hypercube becomes larger and larger as the size of the space interior to the cube increases while the surface to volume ratio of the cube increases linearly with the dimension. This fact has a profound consequence in data analysis since the representation of the data in the high dimensional space becomes very sparse and actually most data fall very close to the surface.

This geometric attraction of the data to the surface in higher dimensions is termed the *curse of dimensionality*. It can be countered by increasing the training data set but this is not always possible. Thus one needs to use dimensionality reduction algorithms that lower the effective dimension of the dataset and bypass this problem; the Principal Component Analysis or PCA is a very popular one. In PCA one uses the variance of the data along different directions and projects the data in a hyperplane that contains most variance. This way the number of space dimensions that is determined by the number of features reduces and the features retained are ones containing most of the essential relationships of the data. This projection into a smaller dimensional space simplifies the problem and makes the results more transparent for analysis and visualization. The first principal component contains most data variance, the second less, the third even less, etc.

In order to apply PCA it is best to start with data standardization, viz. remove mean from each datum while dividing by the standard deviation of the data. Subsequently one forms the covariance matrix that determines the co-dependence of each feature on others. Since this is a square matrix, the method diagonalizes it and produces its eigenvectors that are the principal components and the eigenvalues that determine the contribution to the total variance of each specific component. The hierarchy of the eigenvalues from larger to smaller contains the contribution to the variance of each component. The eigenvector with the largest eigenvalue contains most variance and it is the first principal component, the next eigenvalue determines the second principal component, etc. The data are then projected into the directions with the highest variance so that most variance is retained.

We may perform PCA on the iris data set and find the first two components with the highest variance; the Python code follows while in Fig. 3.1 we show the plot of the data based on the first two PCA components. The explained variance ratio is a metric that indicates how much of the total variance in the data is explained

3.2 Principal Component Analysis

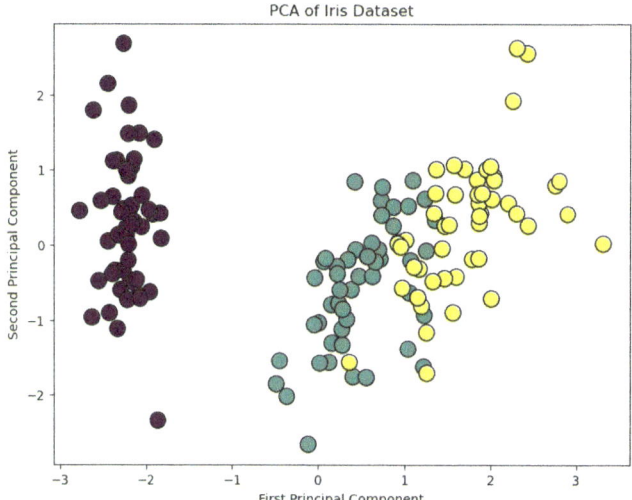

Fig. 3.1 Iris PCA results with explained variance ratio: [0.72962445 0.22850762]. Each cluster with different color corresponds to different iris flower class

by each of the principal components. The first two principal components explain more that 95% of the variance and this leads to almost separable data with only two components.

```
import numpy as np
import matplotlib.pyplot as plt
from sklearn.datasets import load_iris
from sklearn.preprocessing import StandardScaler
from sklearn.decomposition import PCA

# Load the Iris dataset
iris = load_iris()
X = iris.data    # Features
y = iris.target  # Labels

# Standardize the data
scaler = StandardScaler()
X_std = scaler.fit_transform(X)

# Apply PCA (reduce to 2 components)
pca = PCA(n_components=2)
X_pca = pca.fit_transform(X_std)

# Plot the PCA-reduced data
plt.figure(figsize=(8, 6))
plt.scatter(X_pca[:, 0], X_pca[:, 1], c=y, cmap='viridis',
    edgecolor='k', s=150)
plt.xlabel('First Principal Component')
```

```
plt.ylabel('Second Principal Component')
plt.title('PCA of Iris Dataset')
plt.show()

# Explained variance ratio
print("Explained variance ratio:", pca.explained_variance_ratio_)
```

3.2.1 Covariance Matrix

When PCA works on statistical data samples it evaluates first the covariance matrix. Assume that X is and $m \times n$ matrix of m samples and n features, then the covariance matrix K_{XX} is

$$K_{XX} = \frac{1}{m-1} X^T X \tag{3.1}$$

where X^T is the $n \times m$ transpose of X; clearly K_{XX} is a square $n \times n$ matrix with eigenvectors $\upsilon_1, \upsilon_2, \ldots, \upsilon_n$ that are the principle components and $\lambda_1, \lambda_2, \ldots, \lambda_n$ the corresponding eigenvalues. Let us assume that only the first k eigenvectors are kept based on the hierarchy of the eigenvalues and are placed in a new matrix V with dimension $n \times k$. The data projection through PCA is done through

$$X_{PCA} = X \cdot V \tag{3.2}$$

where X_{PCA} is the $m \times k$ reduced projected matrix.

3.3 Singular Value Decomposition

In many cases the diagonalization of a large covariance matrix may pose problems with ease of processing or stability issues. It might then be preferable to decompose the $m \times n$ matrix X of the available data into its singular values and subsequently evaluate the appropriate covariance matrix that will be of lower dimensionality. The *Singular Value Decomposition*, or *SVD*, factorizes any real matrix into a product of three matrices:

$$X = U \Sigma V^T \tag{3.3}$$

where U is an orthogonal $m \times m$ matrix of left singular vectors, Σ is a diagonal $m \times n$ matrix of singular values and V^T is the transpose of an $n \times n$ orthogonal matrix of the right singular vectors. The right singular vectors in column form are

3.3 Singular Value Decomposition

the principal components of the data matrix X while the corresponding singular values express the variance explained by each singular component. It is thus possible after making the SVD decomposition to select those singular vectors that contain the highest variance and proceed with the evaluation of the covariance matrix for this set. after this decomposition. The result should be identical to the one obtained if PCA was evaluated directly. We may re-evaluate the iris data set PCA, however after first performing SVD. The Python code below first performs SVD on the data and then retains the first two principal components. After running the code we obtain the same result as in the previous application of PCA with `Explained variance ratio (first two components): [0.72962445 0.22850762]`.

```python
import numpy as np
import matplotlib.pyplot as plt
from sklearn.datasets import load_iris
from sklearn.preprocessing import StandardScaler

# Load the Iris dataset
iris = load_iris()
X = iris.data    # Features
y = iris.target  # Labels

# Standardize the data (mean 0, variance 1)
scaler = StandardScaler()
X_std = scaler.fit_transform(X)

# Compute SVD of the standardized data matrix
U, S, Vt = np.linalg.svd(X_std)

# Project the data onto the first two principal components (top 2 right
#  singular vectors)
X_pca = X_std @ Vt.T[:, :2]

# Plot the PCA-reduced data
plt.figure(figsize=(8, 6))
plt.scatter(X_pca[:, 0], X_pca[:, 1], c=y, cmap='viridis', edgecolor='k',
    s=150)
plt.xlabel('First Principal Component')
plt.ylabel('Second Principal Component')
plt.title('PCA of Iris Dataset Using SVD')
plt.show()

# Explained variance ratio (Singular values correspond to variance)
explained_variance = (S**2) / np.sum(S**2)
print("Explained variance ratio (first two components):",
    explained_variance[:2])
```

We note that SVD and PCA can be used in synchrony since PCA can use SVD first and find the principal components, i.e. the directions of variance maximization of the data. If we then apply first SVD to the data matrix we may compute the principal components directly without an explicit computation of the covariance matrix in the complete data matrix.

3.4 Gradient Descent

The Gradient Descent method is an optimization algorithm used in machine learning to minimize the *loss or cost function* by iterative updating model parameters in the direction of the steepest descent. The goal is to find the global minimum of the loss function-sometimes even finding a local minimum is useful. The loss function measures objectively how the model's predictions differ from the true values. Let us consider the simpler case of linear regression where the loss function can be expressed as

$$J(\theta) = \frac{1}{2m} \sum_{i=1}^{m} \left(h_\theta(x^{(i)}) - y^{(i)} \right)^2 \qquad (3.4)$$

where m is the number of training examples, $h_\theta(x)$ is the hypothesis or model prediction, $y^{(i)}$ is the actual target value and θ are the weights that are model parameters.

Once a loss function is given then one evaluates the *Gradient*, i.e. the vector partial derivatives of the loss function with respect to the *model parameters*. It is the basic property of the gradient of a function of many variables to point in the direction of steepest increase of the loss function. The gradient of the function is simply the vector of the first partial derivatives with respect to each space direction. For parameter θ_j the partial derivative is

$$\frac{\partial J(\theta)}{\partial \theta_j} \qquad (3.5)$$

As the optimization proceeds in model parameter space and the parameters are updated, there is a basic scalar quantity that controls the step size of this update; this is the *learning rate* α. If the learning rate is too small the convergence to a minimum is slow, if it is too large the algorithm can overshoot it. The learning rate is thus an important hyperparameter in the process of finding minima of the function. The gradient descent algorithm updates the parameters θ iteratively using the formula:

$$\theta_j = \theta_j - \alpha \cdot \frac{\partial J(\theta)}{\partial \theta_j} \qquad (3.6)$$

Where θ_j is the current parameter or weight, α is the learning rate and $\frac{\partial J(\theta)}{\partial \theta_j}$ is the partial derivative of the loss function with respect to θ_j.

3.4 Gradient Descent

There are different ways of applying gradient descent that depend on the data usage and mode of searching. The *Batch Gradient Descent* uses the entire dataset to compute the gradient. This is accurate but relatively slow especially if the dataset is large. The *Stochastic Gradient Descent* or *SGD* updates the parameters using one training example at a time; this saves considerable amount of computer time and the end result is quite good. In *Mini-batch Gradient Descent* a small subset, i.e. a mini-batch of the dataset is used in order to compute the gradient. In the Python code that follows we show a simple implementation of gradient descent for linear regression.

```python
import numpy as np
import matplotlib.pyplot as plt

# Generate some sample data (y = 2x + 3)
np.random.seed(42)
X = 2 * np.random.rand(100, 1)
y = 4 + 3 * X + np.random.randn(100, 1)

# Add bias term (column of ones for the intercept)
X_b = np.c_[np.ones((100, 1)), X]

# Hyperparameters
learning_rate = 0.1
n_iterations = 1000
m = 100  # number of samples

# Initialize theta (weights)
theta = np.random.randn(2, 1)  # two parameters (for intercept and slope)

# Gradient Descent
for iteration in range(n_iterations):
    gradients = 2/m * X_b.T.dot(X_b.dot(theta) - y)
    theta = theta - learning_rate * gradients

# Final parameters (weights)
print("Theta (intercept and slope):", theta)

# Plot the result
plt.plot(X, y, "b.")
plt.plot(X, X_b.dot(theta), "r-", label="Predictions")
plt.xlabel("x")
plt.ylabel("y")
plt.title("Linear Regression with Gradient Descent")
plt.legend()
plt.show()
```

In this example we generate synthetic linear data $y = 4 + 3x + \epsilon$, where ϵ is random noise. A column of ones is added to account for the intercept (bias term) in the linear model. The learning rate is a hyperparameter that controls the size of the

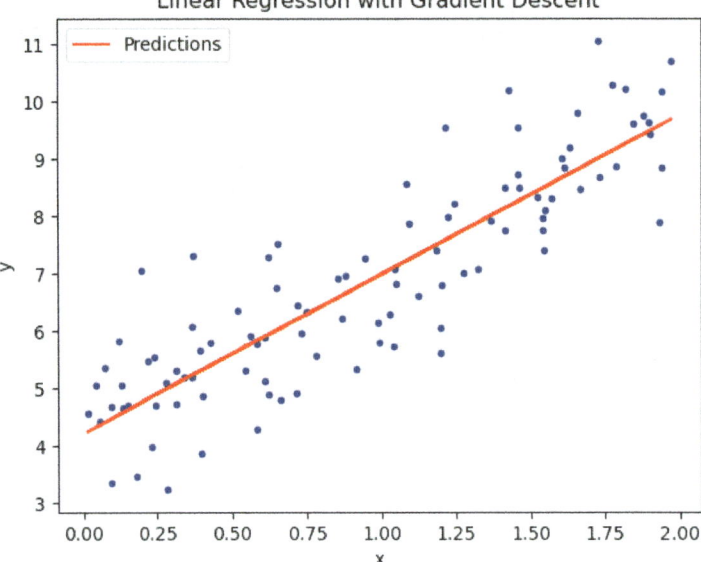

Fig. 3.2 Linear regression with gradient descent. The intercept and slope is [[4.21509616] [2.77011339]] respectively

update steps, and *n_iterations* controls the number of iterations. We initialize the weights θ randomly. In the gradient descent loop, for each iteration, the gradients are computed and the parameters are updated by moving them in the direction of the negative gradient. If α is too large, the algorithm might overshoot the minimum. If α is too small, the algorithm will converge slowly. The algorithm stops either after a fixed number of iterations or when the change in the loss function between iterations is smaller than a defined threshold. The result is shown in Fig. 3.2.

3.5 Visualizing Gradient Descent

To better understand how the gradient descent works we have a Python code below that aids at visualizing the path that gradient descent follows when minimizing a cost function for a two dimensional case.

```
import matplotlib.pyplot as plt
import numpy as np

# Sample 2D quadratic cost function: f(x, y) = (x^2 + y^2)
def cost_function(theta):
    return theta[0]**2 + theta[1]**2
```

```python
# Gradient of the cost function
def gradient(theta):
    return 2 * theta

# Gradient Descent parameters
learning_rate = 0.1
n_iterations = 10
theta = np.array([2.0, 2.0])  # initial point

# Store the points for visualization
path = [theta.copy()]

# Perform Gradient Descent
for iteration in range(n_iterations):
    gradients = gradient(theta)
    theta -= learning_rate * gradients
    path.append(theta.copy())

# Convert path to a numpy array
path = np.array(path)

# Plotting the function and the gradient descent path
X, Y = np.meshgrid(np.linspace(-2.5, 2.5, 100), np.linspace(-2.5, 2.5, 100))
Z = X**2 + Y**2

plt.contour(X, Y, Z, levels=np.logspace(0, 3, 35), cmap='jet')
plt.plot(path[:, 0], path[:, 1], 'ro-', label="Gradient Descent Path")
plt.xlabel("x")
plt.ylabel("y")
plt.title("Gradient Descent Path on Cost Function")
plt.legend()
plt.show()
```

In Fig. 3.3 we see the path that the gradient descent follows in the process of minimization of the cost function.

3.6 Machine Learning Performance Measures

When we apply machine learning it is important to assess the success of the learned methods we use in handling the data. Since we use a certain set of data for training the model it is important to know well the model performs on data not used for training. The evaluation of classifiers as well as for regression is performed through the use of specific measures that give a clear picture on the success or failure of the model.

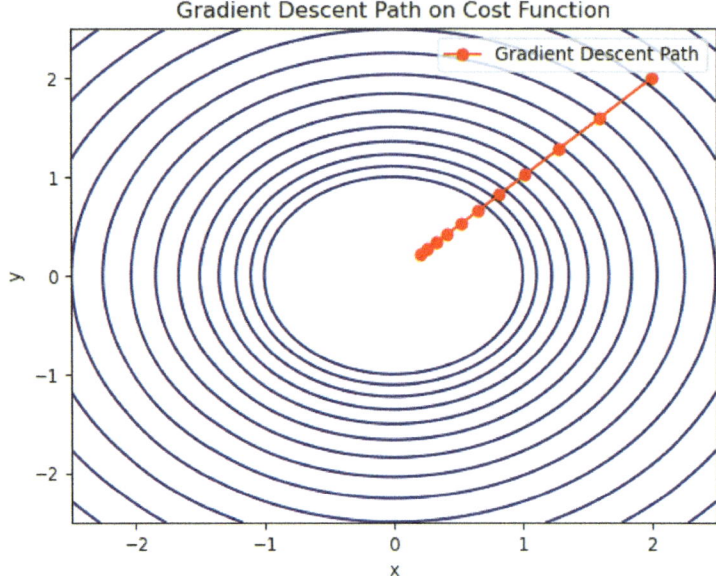

Fig. 3.3 Gradient descent path from a starting point and descending to the minimum of the cost function

3.6.1 Classification Metrics

When we perform a classification task and we have two classes we are typically interested to know the number of instances that one class is incorrectly classified as the other class. This type of information is contained in the *Confusion Matrix* and a number indicators obtained through it. The confusion matrix is a table that shows the actual vs. predicted classifications. It has four elements:

- *True Positives (TP):* Correctly predicted positive instances.
- *True Negatives (TN):* Correctly predicted negative instances.
- *False Positives (FP):* Incorrectly predicted as positive (Type I error).
- *False Negatives (FN):* Incorrectly predicted as negative (Type II error).

The confusion matrix has following form:

	Predicted Positive	Predicted Negative
Actual Positive	TP	FN
Actual Negative	FP	TN

The confusion matrix is useful for getting a complete picture of the performance, especially in multi-class classification. There are a number of measures derived from the confusion matrix. The *Accuracy* is the ratio of correctly predicted instances, to the total number of instances.

3.6 Machine Learning Performance Measures

$$\text{Accuracy} = \frac{\text{Number of correct predictions}}{\text{Total number of predictions}} \quad (3.7)$$

This metric is best used when classes are balanced.

The *Precision* is the ratio of true positive predictions to the total number of positive predictions. The latter includes both the correctly predicted instances, i.e. the true positive ones but also the ones predicted erroneously. The *False Negative* or *FN* instances occur then the model incorrectly predicts the negative class for an instance that actually belongs to the positive class.

$$\text{Precision} = \frac{\text{True Positives}}{\text{True Positives} + \text{False Positives}} \quad (3.8)$$

This metric is useful when the cost of false positives is high.

The *Recall* or *Sensitivity* or *True Positive Rate* is the ratio of true positive predictions to the total number of actual positives.

$$\text{Recall} = \frac{\text{True Positives}}{\text{True Positives} + \text{False Negatives}} \quad (3.9)$$

This metric is important when the cost of false negatives is high, e.g. in medicine in disease detection.

The *F1-Score* is the harmonic mean of precision and recall. It balances the trade-off between precision and recall.

$$F1 = 2 \times \frac{\text{Precision} \times \text{Recall}}{\text{Precision} + \text{Recall}} \quad (3.10)$$

It is useful when you need to balance precision and recall, especially with imbalanced datasets.

The *ROC Curve* provides a plot of the True Positive Rate (Recall) against the False Positive Rate (FPR) for different threshold settings. It is useful for evaluating the trade-offs between true positive rate and false positive rate. The *AUC* or *Area Under the Curve* is the area under the ROC curve. It ranges from 0 to 1, with 1 indicating a perfect classifier. The AUC provides a single number summary of classifier performance, useful for comparing models.

Finally, the *Log Loss* or *Cross-Entropy Loss* measures the performance of a classification model whose output is a probability value between 0 and 1.

$$\text{Log Loss} = -\frac{1}{N} \sum_{i=1}^{N} \left[y_i \log(p_i) + (1 - y_i) \log(1 - p_i) \right] \quad (3.11)$$

where y_i is the actual label and p_i is the predicted probability. This classification measure is useful for models that output probabilities; lower values indicate better performance.

3.6.2 Regression Metrics

In regression the L_1 and L_2 norms are useful metrics leading respectively to *Mean Absolute Error* or MAE that is the average of the absolute differences between the predicted and actual values and the *Mean Squared Error* or MSE that determines the average of the squared differences between the predicted and actual values. We have

$$\text{MAE} = \frac{1}{N} \sum_{i=1}^{N} |y_i - \hat{y}_i| \tag{3.12}$$

where y_i is the actual value and $\hat{y}_i$ is the predicted value.
The MAE gives a straightforward measure of the error in units of the predicted variable. For the MSE

$$\text{MSE} = \frac{1}{N} \sum_{i=1}^{N} (y_i - \hat{y}_i)^2 \tag{3.13}$$

The MSE penalizes larger errors more heavily, making it sensitive to outliers. There is also the related to it *Root Mean Squared Error* or RMSE that simply provides an error measure in the same units as the output variable.

$$\text{RMSE} = \sqrt{\text{MSE}} \tag{3.14}$$

The RMSE is often easier to interpret since it is in the same units as the output and it heavily penalizes large errors.

The *R-Squared* Coefficient of Determination measures the proportion of variance in the dependent variable that is predictable from the independent variables.

$$R^2 = 1 - \frac{\sum_{i=1}^{N}(y_i - \hat{y}_i)^2}{\sum_{i=1}^{N}(y_i - \bar{y})^2} \tag{3.15}$$

where $\bar{y}$ is the mean of the actual values. R^2 provides a measure of how well unseen samples are likely to be predicted by the model. Values range from 0 to 1, with higher values indicating better performance. The *Adjusted R-Squared* is a modified version of R^2 that adjusts for the number of predictors in the model.

$$\text{Adjusted } R^2 = 1 - \left(\frac{1 - R^2}{n - p - 1} \right) \times (n - 1) \tag{3.16}$$

where n is the number of observations and p is the number of predictors.

3.7 Conclusions

The adjusted R^2 is useful when comparing models with different numbers of predictors.

3.6.3 Other Metrics

The *Silhouette Score* is a clustering metric and measures how similar an object is to its own cluster compared to other clusters.

$$s(i) = \frac{b(i) - a(i)}{\max(a(i), b(i))} \quad (3.17)$$

where $a(i)$ is the average distance between i and all other points in the same cluster, and $b(i)$ is the average distance between i and all points in the nearest cluster. The silhouette score ranges from -1 to 1, with higher values indicating better-defined clusters.

Finally the *Cross-Validation Score* is quite important for cross-validation where the model is trained and evaluated on different subsets of the data multiple times. The cross-validation score is typically the average performance across all folds. It is quite useful for evaluating model robustness and avoiding overfitting. The `cross_val_scores()` function of `scikit-learn` performs this task and can perform a k-fold cross-validation that involves splitting the dataset into k subsets. The model is trained on $k-1$ of these folds and tested on the remaining one. This process is repeated k times, each time with a different fold as the test set.

3.7 Conclusions

The curse of dimensionality can be cured if we apply appropriate methods in the dataset. The PCA that diagonalizes the covariance matrix gives the important directions in high dimensional spaces that contain most data relevant information. It is a very important method and it is used routinely in AI. It might be preferable to use SVD to the data directly and obtain the singular values that dominate their distribution. Use of the most important singular vectors reduces the effective dimensionality of the data and makes their analysis much simpler. The gradient descent is the method of choice in machine learning for finding the optimal parameters in a model during training. In order to see how well the trained model generalizes we have an arsenal of metrics both for classification and regression. In the former the confusion matrix contains essential information on how well the model classifies. In regression metrics such as mean absolute error or mean square error give the corresponding information. Cross-validation is a very important method that give information on how well a model generalizes or overfits. Clustering methods such as the silhouette score and others determine how coherent are clusters of data in terms of describing the same features. The metrics presented in the chapter are used in the machine learning models in order to see how well the model performs and are also used in determining optimal hyperparameters.

3.8 Summary

- Principal components analysis or PCA determines space directions with the higher variance.
- Singular value decomposition of SVM finds a hierarchy of singular values with singular vectors in order of importance.
- Gradient descent methods provide the optimal parameters for a machine learning model.
- The confusion matrix determines how well the model classifies data
- In regression we use mean absolute or mean square error or other functions to measure how well the model predicts.

References

1. A. Géron, *Hands-on Machine Learning with Scikit-Learn, Keras and TensorFlow*, 3rd edn. (O'Reilly Media, Inc., 2023)
2. I. Goodfellow, Y. Bengio, A. Courville, *Deep Learning* (MIT Press, 2016)
3. O. Campesato, *Artificial Intelligence, Machine Learning and Deep Learning* (Mercury Learning and Information, Dulles, 2020)

Part II
Artificial Neural Networks and Deep Learning

Chapter 4
Artificial Neurons and Deep Learning

The Extreme Power of Neurons

Abstract Real biological neurons form networks through synapses and learn. In a similar way their artificial counterparts form networks that can be trained to perform certain tasks. The neuron involves inputs and outputs and a central nonlinear part that helps selection of possibilities. Deep neural networks with several hidden layers are able to perform complex tasks that simpler neuron connections cannot. Artificial intelligence grew out of the construction and study of artificial neural networks.

4.1 Artificial Intelligence

Artificial intelligence or AI is closely linked to the development of the ideas of neural networks. During the 1940s Warren McCulloch and Walter Pitts constructed the first mathematical model for a neuron, called appropriately the *McCulloch-Pitts neuron* [1]. This model has a binary output and aimed at mimicking the function of biological neurons. This important idea led in the 1950s and 60s to the *perceptron*, a simple neural network introduced by Frank Rosenblatt and that is capable of classifying inputs in two categories [2]. The input values summed with appropriate weights *W* and the result is filtered through a nonlinear threshold-type function to produce and output, as in Fig. 4.1. Even through the introduction of the perceptron was a powerful idea for the period, Marvin Minsky and Seymur Papert in the late 1960s showed that it had serious limitations [3]. For instance, it could not produce certain functions such as the XOR logical function. This lead to a decline in the interest in neural networks until the 1980s when two important results appeared. In 1982 John Hopfield introduced a neural network model for associative memory that was based on simple Ising-type statistical mechanics models with random interactions [4]. This idea lead to a furry of both experimental [5] and theoretical activity [6, 7] in the area since it connected with well-known problems of statistical physics. The second idea was similarly very significant since it introduced a method for systematic evaluation of the *artificial neural network* or *ANN* weights. This is the algorithm of *backpropagation* by David Rumelhart, Geofrey Hinton and Ronald J. Williams [8]. In this algorithm the ANN learns by adjusting its weights based on the error of the output. Backpropagation permitted efficient learning in more

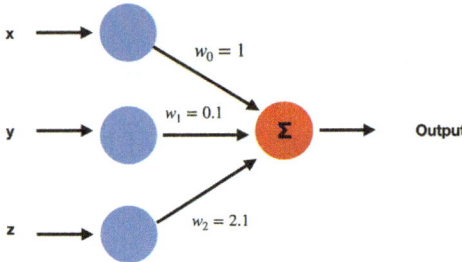

Fig. 4.1 A Perceptron has inputs and a single binary output. The weights W determine how the inputs are passed through the nonlinear element that determines the binary output

complex multi-layer perceptrons through weight adjustment and error correction. More advances occurred in the 1990s such as the introduction of recurrent neural networks (RNNs), convolutional neural networks (CNNs), etc but with limited use due to lack of appropriate computing power. In 2000s and 2010s the rise of GPUs and large data sets led to the development of *deep learning* that consists of neural networks with many hidden layers [9]. These developments now drive the explosive use of AI in various areas from healthcare to education to self-driving technologies.

4.2 The OR and XOR Neural Networks

In order to fix ideas and see how artificial neural networks are applied we will use a simple example related to the logical gates OR and XOR. The truth table for the two gates is given in Table 4.1.

The difference between the OR and XOR gates is only in the last line of the truth table; when both A and B are true while the OR gate is true the XOR is false. While the OR gate is simple to code the same is not true for the XOR gate; the latter is not linearly separable. This feature is easier shown in a two dimensional diagram as in Fig. 4.2. In both cases we have two classes designated by 0 and 1. In the OR gate the input pair $(0, 0)$ belongs to the first class while the other three input pairs $(1, 0)$, $(0, 1)$, $(1, 1)$ to the second class. We can draw a line on the plane that clearly separates the two classes. In the XOR gate both $(0, 0)$ and $(1, 1)$ pairs correspond to the same class; thus there is no line that splits the classes according to the values of the input pairs.

A Python code for the OR gate follows.

```
import tensorflow as tf
from tensorflow.keras.models import Sequential
from tensorflow.keras.layers import Dense
import numpy as np

# Define the OR gate inputs and outputs
inputs = np.array([[0, 0], [0, 1], [1, 0], [1, 1]])
outputs = np.array([[0], [1], [1], [1]])
```

4.2 The OR and XOR Neural Networks

Table 4.1 Truth Tables for OR ($A \vee B$) and XOR ($A \oplus B$) gates

Inputs		OR	XOR
A	B	$A \vee B$	$A \oplus B$
0	0	0	0
0	1	1	1
1	0	1	1
1	1	1	0

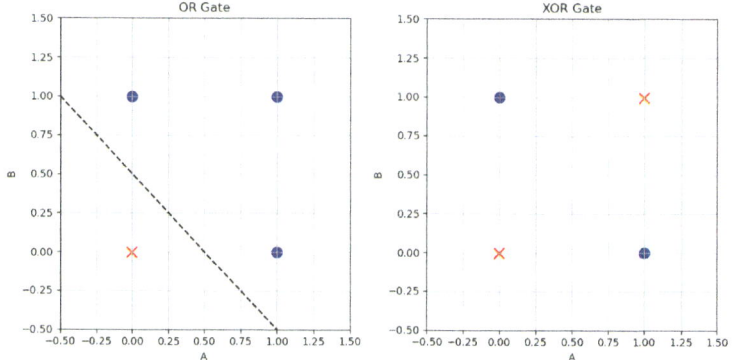

Fig. 4.2 Linear separability in two classes of the OR gate, non-separability for the XOR gate

```
# Build the neural network model
model = Sequential()
model.add(Dense(1, input_dim=2, activation='sigmoid'))

# Compile the model
model.compile(loss='binary_crossentropy', optimizer='adam',
    metrics=['accuracy'])

# Train the model
model.fit(inputs, outputs, epochs=1000, verbose=0)

# Test the model
predictions = model.predict(inputs)
print('Predictions:')
print(predictions.round())  # Round the output to 0 or 1
```

while the Python code for the XOR gate is bellow here

```
import tensorflow as tf
from tensorflow.keras.models import Sequential
from tensorflow.keras.layers import Dense
import numpy as np
```

```
# Define the XOR gate inputs and outputs
inputs = np.array([[0, 0], [0, 1], [1, 0], [1, 1]])
outputs = np.array([[0], [1], [1], [0]])

# Build the neural network model
model = Sequential()
model.add(Dense(2, input_dim=2, activation='relu'))
# Hidden layer with 2 neurons
model.add(Dense(1, activation='sigmoid'))  # Output layer

# Compile the model
model.compile(loss='binary_crossentropy', optimizer='adam',
    metrics=['accuracy'])

# Train the model
model.fit(inputs, outputs, epochs=1000, verbose=0)

# Test the model
predictions = model.predict(inputs)
print('Predictions:')
print(predictions.round())  # Round the output to 0 or 1
```

In both codes we introduce artificial neural networks through the use of **TensorFlow** and **Keras** for instructive reasons. These platforms are basic for AI and their existence led in part to the widespread circulation and applications of machine learning. A short introduction on their use can be found in Appendix A where Python-related coding is briefly described. There is a major difference between the two codes reflecting the fact that the XOR gate is not linearly separable. In the OR gate case all we need is a single neuron that is fully capable to make the separation in the two classes. This neuron has two input values for A and B respectively and one output value.

Let us now zoom in the neuron and see its internal structure starting from the OR gate. Assume that the internal variables for the A and B input nodes are respectively x_1 and x_2 and they take values 0 or 1. These values are multiplied with the appropriate weights W_1 and W_2 and the result is added up and together with the bias term b enters in the central node of the neuron as in Fig. 4.3. The linear function

$$z = W_1 x_1 + W_2 x_2 + b \qquad (4.1)$$

determines then the input to the nonlinear node of the neuron. The weights W_1, W_2, b are the only unknown parameters in the neuron and are determined through *training* that will be explained shortly. Once the weighted sum z is formed, its value passes through the activation function. This is a nonlinear function that typically has a threshold-like structure and is the basic nonlinear function of the neuron. Typical activation functions are the *sigmoid* that maps the output to a range between 0 and 1, the *tanh* that maps it in the range -1 to 1, the *ReLU* (Rectified Linear Unit) that outputs directly the input if it is positive or else outputs 0 as well

4.2 The OR and XOR Neural Networks

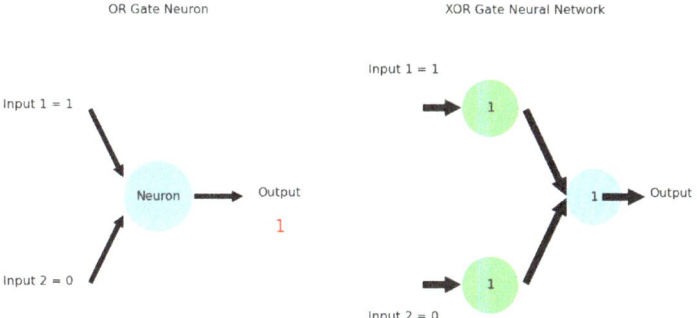

Fig. 4.3 Visualization of the OR and XOR gates

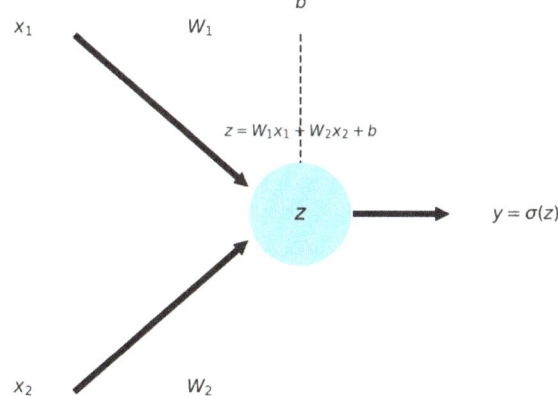

Fig. 4.4 OR gate with one neuron and three unknown weights. The activation function $\sigma(z)$ is sigmoid

as other more complex functions. In the OR gate we used the sigmoid function. The exit value from the activation function is the neuron output.

The artificial neuron structure then consists of the input and output values that is completely dominated by the data as well as its "internal" structure that consists of the unknown weights and the selection of the appropriate activation function. Once the weights W_1, W_2, b are determined, the neuron can act autonomously and hopefully provide the function it was designed for. This function is accomplished through training, viz. through the use of *known* input-output pairs of data that are used to determine the weights. In this iterative procedure the goal is to find the weights that best connect the input and output pairs of data in the training set. This is done through the minimization of an error or cost function that compares true data output with the output of the neuron. This iteration ends when the distance between real and produced data is minimal. Here we choose to perform a fixed number of 1000 iterations. Usual loss functions include the mean square error, but also more complex one such as the *cross entropy* function that is used in the code for the OR gate. The pictorial representation of the single neuron OR gate is in Fig. 4.4.

The XOR gate is more complex since it is not linearly separable as mentioned earlier and it cannot be simulated by just a single neuron. In the Python code we used in addition to the input variables and the output neuron as in the OR gate a *hidden layer* consisting of two neurons. Thus, in total for the XOR gate we use three neurons, two of which form an intermediate layer; for these two neurons we choose ReLU as the activation function while sigmoid is still the preferred choice for the output neuron. The addition of the hidden layer of two neurons adds substantially in the complexity of the problem. Let us describe mathematically the XOR in more detail.

The input layer consists of two input neurons x_1 and x_2, then a hidden layer with two more neuron and finally in the output layer one neuron. We need to use appropriate labeling for the input of weights and biases for each layer.

For the first layer the weights that are input to hidden layer are:

$W_1^{(1)}, W_2^{(1)}$: Weights connecting x_1 and x_2 to the first hidden neuron.

$W_1^{(2)}, W_2^{(2)}$: Weights connecting x_1 and x_2 to the second hidden neuron.

For the second layer weights that connect the hidden layer to the output we have:

$W_1^{(3)}, W_2^{(3)}$: Weights connecting the hidden neurons to the output neuron.

b_1 : Bias for the first hidden neuron, b_2 : Bias for the second hidden neuron,

and b_{out} : Bias for the output neuron.

The two inputs, x_1 and x_2, are passed through the first layer, which consists of two hidden neurons. The hidden neurons perform a weighted sum of the inputs plus a bias and then apply an activation function.

First Hidden Neuron

$$z_1 = W_1^{(1)} \cdot x_1 + W_2^{(1)} \cdot x_2 + b_1$$

The output of the first hidden neuron is the result of applying the activation function $\sigma(z)$, typically a sigmoid function or ReLU:

$$h_1 = \sigma(z_1)$$

For the sigmoid activation function:

$$\sigma(z_1) = \frac{1}{1 + e^{-z_1}}$$

Second Hidden Neuron

$$z_2 = W_1^{(2)} \cdot x_1 + W_2^{(2)} \cdot x_2 + b_2$$

The output of the second hidden neuron is:

$$h_2 = \sigma(z_2)$$

Again, if the activation function is sigmoid:

$$\sigma(z_2) = \frac{1}{1+e^{-z_2}}$$

2. Output Layer Computation

The outputs of the two hidden neurons, h_1 and h_2, are passed to the output layer. The output neuron performs a weighted sum of the hidden neurons' outputs plus a bias, followed by the activation function.

$$z_{out} = W_1^{(3)} \cdot h_1 + W_2^{(3)} \cdot h_2 + b_{out}$$

The output of the output neuron is:

$$y = \sigma(z_{out})$$

For the sigmoid activation function:

$$y = \frac{1}{1+e^{-z_{out}}}$$

For an XOR gate, the goal is for the output y to be 1 when the inputs x_1 and x_2 are different, and 0 when they are the same. This is achieved by learning appropriate weights $W_1^{(1)}, W_2^{(1)}, W_1^{(2)}, W_2^{(2)}, W_1^{(3)}, W_2^{(3)}$ and biases b_1, b_2, b_{out} through training. These nine parameters that need to be fixed through the training process. This a considerable larger number than in the simpler OR gate. We note that the training is a critical and time-consuming process that is however very important for the creation of a faithful model for the function at hand, i.e. the XOR gate in the specific case (Fig. 4.5).

We know that in machine learning once the training is accomplished we need to use the test set of data in order to test how well the artificial system simulates the function we aim for. In the code below we use a single Python code that trains both gates and then in a separate code we evaluate the success of the models.

```python
import tensorflow as tf
from tensorflow.keras.models import Sequential
from tensorflow.keras.layers import Dense
import numpy as np

# OR Gate
# Define the OR gate inputs and outputs
inputs_or = np.array([[0, 0], [0, 1], [1, 0], [1, 1]])
outputs_or = np.array([[0], [1], [1], [1]])

# Build the OR gate model
model_or = Sequential()
model_or.add(Dense(1, input_dim=2, activation='sigmoid'))

# Compile the model
model_or.compile(loss='binary_crossentropy', optimizer='adam',
    metrics=['accuracy'])

# Train the model
model_or.fit(inputs_or, outputs_or, epochs=1000, verbose=0)

# Save the OR gate model
model_or.save('or_gate_model.h5')

# XOR Gate
# Define the XOR gate inputs and outputs
inputs_xor = np.array([[0, 0], [0, 1], [1, 0], [1, 1]])
outputs_xor = np.array([[0], [1], [1], [0]])

# Build the XOR gate model
model_xor = Sequential()
model_xor.add(Dense(2, input_dim=2, activation='relu'))  # Hidden layer
 #  with 2 neurons
model_xor.add(Dense(1, activation='sigmoid'))  # Output layer

# Compile the model
model_xor.compile(loss='binary_crossentropy', optimizer='adam',
    metrics=['accuracy'])

# Train the model
model_xor.fit(inputs_xor, outputs_xor, epochs=1000, verbose=0)

# Save the XOR gate model
model_xor.save('xor_gate_model.h5')
```

4.2 The OR and XOR Neural Networks

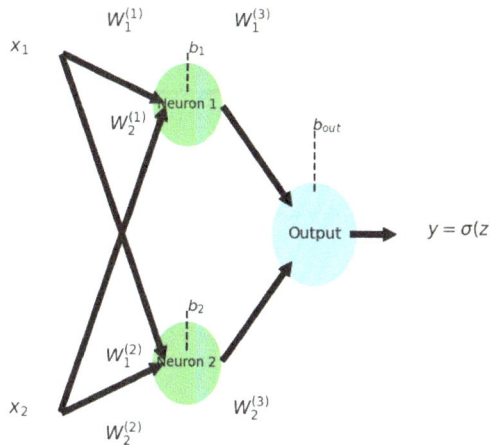

Fig. 4.5 XOR gate with one intermediate layer of two neurons and one output neuron. Here we take all activation functions the same but this is not necessary. In the Python code we se the first two neurons to have ReLU activation and in the output neuron to be sigmoid

And for the test

```
import tensorflow as tf
from tensorflow.keras.models import load_model
import numpy as np

# Load the saved OR and XOR gate models
model_or = load_model('or_gate_model.h5')
model_xor = load_model('xor_gate_model.h5')

# Define the test inputs
test_inputs = np.array([[0, 0], [0, 1], [1, 0], [1, 1]])

# Expected outputs (for verification)
expected_outputs_or = np.array([[0], [1], [1], [1]])
expected_outputs_xor = np.array([[0], [1], [1], [0]])

# Predict using the loaded OR gate model
print("Testing OR Gate Model")
predictions_or = model_or.predict(test_inputs)
print("Test Inputs:\n", test_inputs)
print("Predicted Outputs:\n", predictions_or.round())
print("Expected Outputs:\n", expected_outputs_or)
accuracy_or = np.mean(predictions_or.round() == expected_outputs_or)
print(f"OR Gate Model Accuracy: {accuracy_or * 100:.2f}%\n")

# Predict using the loaded XOR gate model
print("Testing XOR Gate Model")
predictions_xor = model_xor.predict(test_inputs)
print("Test Inputs:\n", test_inputs)
print("Predicted Outputs:\n", predictions_xor.round())
print("Expected Outputs:\n", expected_outputs_xor)
```

```
accuracy_xor = np.mean(predictions_xor.round() == expected_outputs_xor)
print(f"XOR Gate Model Accuracy: {accuracy_xor * 100:.2f}%\n")
```

After running the codes we find that the accuracy for both models is 75% that is not very high. This means that our models can be improved by making various changes in terms the choices of neurons and activation functions as well as the training part.

4.3 Deep Learning

The presence of a hidden layer in the AI code is very useful and versatile and can solve problems that cannot be handled in its absence. Can we put more hidden layers between input and output? Certainly and this leads to *deep networks*, i.e. neural networks that have at least two hidden layers, as in Fig. 4.6. Each layer is independent of the other, contains as many neurons we deem appropriate and actually has a activation function of our choice. We refer as *width* the number of nodes in a layer and as *depth* the number of layers in the network. As the width and depth increases so does the number of training parameters in the neural network. This in turn demands more computer time and very importantly a substantial number of training data. The process of backpropagation becomes quite involved and numerous techniques are employed in order to make it more attainable. If the weights are chosen randomly initially, then it takes considerable time for backpropagation to find the necessary minima. One may opt to use the weights from a *pretrained model* that is related to the problem at hand; this will make the process of training faster and more efficient. The role of hyperparameters is also important and specific values are usually obtained after extended experimentation and empirical knowledge.

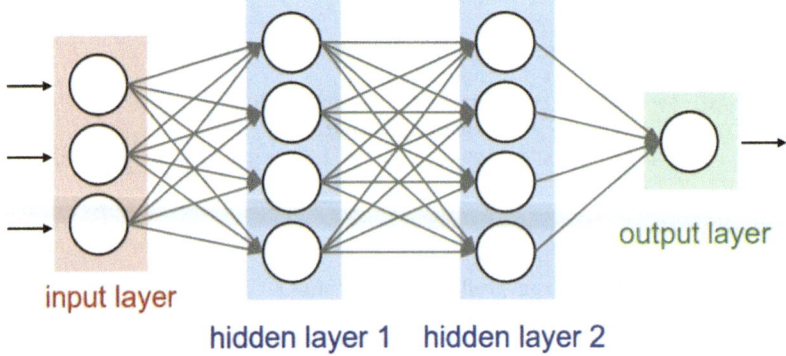

Fig. 4.6 Deep neural network architecture

4.3.1 Overfitting and Regularization

As the neural network model becomes more complex the problem of overfitting becomes more acute. We detect overfitting when the performance of the model in the training set of data is far superior to the one in the test set. When this happens the model cannot generalize to unknown cases and cannot be really trusted due to overfitting. Ways to bypass this problem include the simplification of the model so that it contains fewer fitting parameters that will be able to train with the available data. Generally, the source of overfitting is a disparity between the number of available data and the number of weights to be trained in the neural network model. If more data can be obtained then the model will behave better.

A good method to fight overfitting is through the introduction of constraints in the data function. This is a *regularization* of the data and in many complex cases helps substantially in making the model generalize better. The most usual methods for this type of regularization include L_1, L_2 and *Elastic Net* regularization. The L_1 or Lasso regularization adds a penalty in the loss function that depends on the absolute value of the weight coefficients W_i of the model, viz.

$$L_1 = \text{Original Loss} + \lambda \sum_{i=1}^{n} |W_i| \qquad (4.2)$$

where λ is the regularization parameter that controls how strong is the penalty. In the Lasso regularization the weights enter through their absolute vales and this promotes sparsity of weights and drives some of them to zero. This means that the final model is simpler than the original one without regularization. The L_2 or Ridge regularization adds a penalty that is proportional to the square of the weights. This feature shrinks the values of weights without however reducing them to zero as in L_1 regularization. The new loss function is

$$L_2 = \text{Original Loss} + \lambda \sum_{i=1}^{n} W_i^2 \qquad (4.3)$$

where λ is again the penalty introduced in the model. Finally the Elastic net regularization tries to balance between Lasso and Ridge:

$$L_{EN} = \text{Original Loss} + \lambda_1 \sum_{i=1}^{n} |w_i| + \lambda_2 \sum_{i=1}^{n} w_i^2 \qquad (4.4)$$

where λ_1, λ_2 are the hyperparameters that control the L_1 and L_2 regularizations respectively.

There also other methods of regularization that may be used depending on the problem at hand. *Dropout* is a popular technique where, during each training iteration, a random subset of neurons is "dropped out" or ignored. This reshuffles the neuron contributions to the output and makes the network less reliable on particular neurons. The dropout is applied in the training but not when the network is actually used. In *Early stopping* the model's performance is monitored on a validation set during training. When the performance on the validation set stops improving then the training process stops. This method reduces overfitting by reducing the training time. *Data augmentation* involves artificial increase of the size and variability of the training dataset through the application of random transformations such as rotations, translations, or flipping to the input data. This helps generalization through exposition of the model to a larger dataset. Through *Batch Normalization* we normalize the inputs to each layer and increase the learning rates; this also helps prevent overfitting. Finally through *Max-Norm Regularization* one constrains the magnitude of the weights and enforces an upper bound in the norm of the weights. This prevents the weights from growing too large and thus controls overfitting as well.

4.4 Conclusions

After approximately eighty years from the first idea on artificial neurons, artificial intelligence uses them now very efficiently to make networks of neurons with very powerful properties. The use of hidden layers of neurons leads to deep learning, a mode that is able to handle large sets of training data and generalize to the unknown cases. The basic tool for training the weights of the network is backpropagation, i.e. iterate through all weights in such a way as to minimize the distance in the output between data and predictions. Once we select a specific neural network the only free parameters to train are the network weights. The selection however of the model involves various other decisions, such as the depth of the network, the width of its layers, the type of the activation functions to use, etc. Additionally, there are typically several hyperparameters that need to be selected, including the learning rate, possible regularization etc. Once all these aspects are controlled the neural network can generalize and produce very useful information on new data sets. This is in reality the role of AI, much as natural intelligence, to use existing information and knowledge in order to propel experience to unknown territories.

4.5 Summary

- Neurons receive inputs from the input or other neurons and processes them through a typically nonlinear activation function to other neurons or the output
- The OR gate is linearly separable and models through a single neuron but the XOR gate needs an additional layer of two more neurons to function correctly
- Deep learning involves more than one hidden layer of neurons

References

1. W.S. McCulloch, W.H. Pitts, A logical calculus of the ideas immanent in nervous activity. Bull. Math. Biophys. **5**, 115–133 (1943)
2. F. Rosenblatt, The perceptron: A probabilistic model for information storage and organization in the brain. Psychol. Rev. **65**(6), 386–408 (1958)
3. M. Minsky, S. Papert, *Perceptrons: An Introduction to Computational Geometry* (MIT Press, Cambridge, MA, USA, 1969)
4. J.J. Hopfield, Neural networks and physical systems with emergent collective computational abilities. Proc. Natl. Acad. Sci. **79**(8), 2554–2558 (1982)
5. D. Psaltis, A. Sideris, A.A. Yamamura, A multilayered neural network controller. IEEE Control Syst. Mag. **8**(2), 17–21 (1988)
6. A. Engel, C. Van den Broeck, *Statistical Mechanics of Learning* (Cambridge University Press, 2001)
7. D.J.C. MacKay, *Hopfield Networks*. Information Theory, Inference and Learning Algorithms (Cambridge University Press, 2003)
8. D. Rumelhart, G. Hinton, R. Williams, Learning representations by back-propagating errors. Nature **323**, 533–536 (1986)
9. I. Goodfellow, Y. Bengio, A. Courville, *Deep Learning* (MIT Press, 2016)

Chapter 5
Powerful Neural Network Architectures
CNNs and RNNs

Abstract Convolutional neural networks have strong capabilities in image recognition and typically use coarse-graining schemes that reduce the numbers of pixels in the image. They perform discrete convolutions accompanied by information reduction and can be very successful in probabilistic classification. The recurrent neural networks in addition to passing forward information they retain information from previous steps. This makes them particularly useful in time-series predictions. The long short-term memory network has a cell that can retain both short and longer term memory.

5.1 Introduction

In this chapter we focus on more complex and sophisticated neural network architectures. The Convolutional Neural Networks (CNNs) are particularly well-suited for tasks involving grid-like data, such as images or time series. They are a type of deep neural network designed to automatically and adaptively learn spatial hierarchies of features from input data. CNNs are widely used for image classification, where they can categorize images into different groups (e.g., classifying images of animals into categories like cats, dogs, or birds). In addition, CNNs are highly effective in object detection tasks, allowing for the identification and localization of multiple objects within an image. For example, a CNN can identify objects in an image and provide bounding boxes around them. While CNNs are primarily known for image-based tasks, they are also used in some natural language processing (NLP) tasks, such as text classification and sentiment analysis, by capturing patterns in sequences of words or characters. Moreover, CNNs are applied in video analysis tasks like action recognition, object tracking, and activity detection by analyzing frames over time, making them suitable for tasks requiring temporal understanding. They are also widely used in medical image analysis for detecting anomalies like cancerous regions in X-rays, MRIs, and CT scans.

The Recurrent Neural Networks (RNNs) are a type of neural network designed for processing sequential data, where the current input depends on previous inputs. Unlike feed forward networks, RNNs have connections that form cycles, allowing information to persist across time steps. This makes RNNs well-suited for tasks involving time-series data, speech recognition, natural language processing (NLP), and tasks where the order of the data matters. In NLP, RNNs are used for tasks such as language modeling, machine translation, and text generation, as they can capture the context and dependencies between words in a sentence. In time-series analysis, RNNs are employed for forecasting and anomaly detection, leveraging their ability to recognize patterns over time. Speech recognition systems use RNNs to convert spoken language into text by learning the temporal relationships between audio signals. Despite their ability to capture sequential dependencies, RNNs suffer from issues like vanishing gradients, which limit their effectiveness in modeling long-term dependencies. To address these limitations, variants like Long Short-Term Memory (LSTM) and Gated Recurrent Units (GRU) are commonly used to improve their performance on complex sequential tasks.

5.2 Convolutional Neural Networks

Mathematically a convolution is an operation that involves two functions $f(t)$ and $g(t)$ of the independent variable t and produces a third function usually denoted as $(f * g)(t)$ and defined as

$$(f * g)(t) = \int_{-\infty}^{+\infty} f(\tau)g(t - \tau)d\tau \tag{5.1}$$

Intuitively, we may think of sliding one function, say the function $g(t)$, over the function $f(t)$ and compute the integral of their product in each position. If $f(t)$ is a signal then $g(t)$ acts as a filter that combines information accumulated on the scale of the function $g(t)$. In convolutional neural networks or CNNs a similar in spirit discrete convolution is applied in order to filter and reduce the amount of information contained in the original signal. It is a deep learning method that is particularly suited to image classification and object recognition as well as other applications in natural language processing (NLP), audio signals etc. The CNNs focus on spatial structures and learn hierarchical features ranging from simple edges to more complex shapes.

In TensorFlow the convolutional layer is conv2d in two dimensions[1]; let us now see how this function works and what it does.

---

[1] For one-dimensional data we use conv1d and in three dimensions conv3d.

5.2 Convolutional Neural Networks

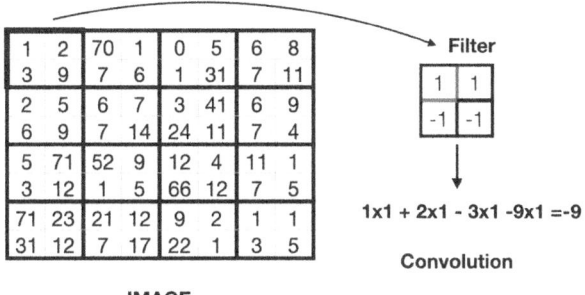

Fig. 5.1 Two dimensional discrete convolutions with a 2 × 2 filter. The result is negative

5.2.1 Convolutional Layer

The main operation of the convolutional layer is performed through a small matrix called *filer* or *kernel*. The filter slides over input data, for instance over an image, and performs element-wise multiplication followed by summation. This operation extracts gross features like edges, textures, or patterns from the input data. In Fig. 5.1 we display a 8 × 8 grid of numbers that plays the role of an image together with a small 2 × 2 filter. The inner product of the first 2 × 2 submatrix with the 2 × 2 filter gives the result − 9. This result appears in the feature map that each filter produces and highlights the presence of specific features in different regions of the input.

The step size by which each filter moves over the input is called *stride*. A stride of 1 moves the filter one pixel at a time, while a larger stride skips pixels, resulting in a smaller output. Furthermore, it might be necessary to add extra pixels that are often zeros around the input to ensure that the filter covers the edges of the input. This process is called *padding* and ensures control of the size of the output feature map.

5.2.2 Activation Function

After the feature maps are created it is possible that some values are negative. The use of ReLU activation function replaces possible negative values with zero. The definition of ReLU is

$$\text{ReLU}(x) = \begin{cases} 0 & \text{if } x \leq 0, \\ x & \text{if } x > 0. \end{cases} \qquad (5.2)$$

Other activation functions as sigmoid or tanh are also possible but RElU is by far the most popular one in CNNs due to is simplicity and efficiency.

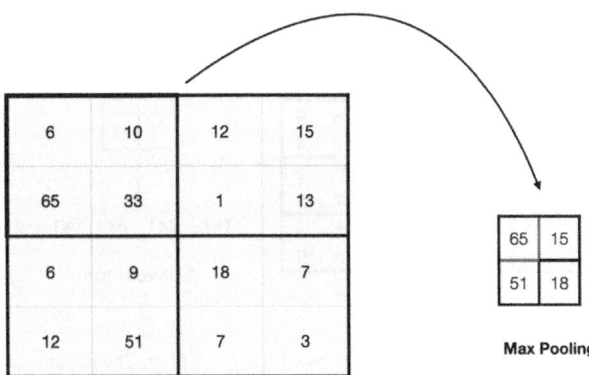

Fig. 5.2 Max Pooling in a CNNs enforces a course graining. In this example it selects the largest value in each 2 × 2 square unit

5.2.3 Max Pooling Layer

The pooling layers reduce the spatial dimensions of the feature maps while retaining the most important information. The *max pooling* step is very simple to perform and involves the partition of the updated feature map in 2 shapes and then select the largest value as the one representing this rectangle. This projection-like process discards 75% of the data in the feature map. One could alternatively select the average number in the 2 × 2 square, but this method is not as popular as simply selecting the largest value. An example of this processes is shown in Fig. 5.2.

5.2.4 Flattening and Softmax

The process of using a convolutional layer that is followed by a max pooling layer can be repeated several times. When this process ends, one usually places one or more fully connected layers that are waiting to processes the last image but in a flattened, long vector form. These last layers are similar to a usual neural networks, are fully connected and lead to the output layer. In classification tasks, the very final layer is usually a softmax[2] layer converting the output into probabilities for each

---

[2] The softmax function, commonly used in machine learning for multi-class classification problems, converts a vector of raw scores or logits into a probability distribution, where the probabilities of each class sum to 1. Given a vector of raw scores $z = [z_1, z_2, \ldots, z_n]$, the softmax function for the i-th element is defined as:

$$\text{softmax}(z_i) = \frac{e^{z_i}}{\sum_{j=1}^{n} e^{z_j}}$$

5.2 Convolutional Neural Networks

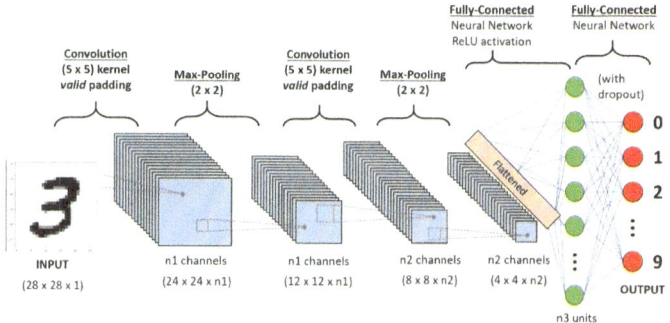

Fig. 5.3 Example of a CNN used for handwritten digit recognition in the MNIST database. The digits have 28 × 28 pixels and pass through a CNN and a Max Pooling layer twice before they go through a fully connected neural network and then another one in the output. The processes aims at recognizing single digits and thus the output has 10 nodes

Fig. 5.4 Handwritten image of 2 from the MNIST database

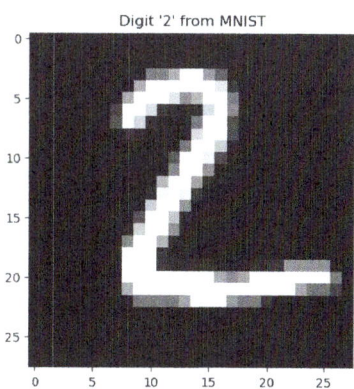

class, making it easier to interpret the model's predictions. The complete process of the CNN is shown in Fig. 5.3 where it is applied to the MNIST database used to recognize handwritten digits.

The CNNs learn hierarchical features automatically; in the early layers they learn simple features such as edges, gradients, and textures. In middle layers they learn more complex features like shapes, corners, and parts of objects while in deeper layers they combine complex features in order to recognize objects, faces, or other high-level patterns in the input data.

The code below opens the MNIST database and simply prints the image of the number 2. The output is shown in Fig. 5.4.

```
import tensorflow as tf
import numpy as np
import matplotlib.pyplot as plt

# Load the MNIST dataset
```

```python
mnist = tf.keras.datasets.mnist
(x_train, y_train), (x_test, y_test) = mnist.load_data()

# Find and display an image of the digit "2" from the test set
index = np.where(y_test == 2)[0][0]  # Find the first instance of the
                                     # digit "2"
image_of_2 = x_test[index]

# Display the image
plt.imshow(image_of_2, cmap='gray')
plt.title("Digit '2' from MNIST")
plt.show()
```

A tensorflow based Python code that uses CNNs in the MNIST database is shown below.

```python
import tensorflow as tf

# Load the MNIST dataset
mnist = tf.keras.datasets.mnist
(x_train, y_train), (x_test, y_test) = mnist.load_data()

# Normalize the data to a range of 0 to 1
x_train, x_test = x_train / 255.0, x_test / 255.0

# Reshape the data to include the channel dimension (necessary for CNNs)
x_train = x_train.reshape(-1, 28, 28, 1)
x_test = x_test.reshape(-1, 28, 28, 1)

# Build the CNN model
model = tf.keras.models.Sequential([
    tf.keras.layers.Conv2D(32, (3, 3), activation='relu',
        input_shape=(28, 28, 1)),
    # 32 filters, 3x3 kernel
    tf.keras.layers.MaxPooling2D((2, 2)),   # Max pooling with 2x2 pool
    #   size
    tf.keras.layers.Conv2D(64, (3, 3), activation='relu'),   # 64 filters,
    #   3x3 kernel
    tf.keras.layers.MaxPooling2D((2, 2)),   # Max pooling with 2x2 pool
    #   size
    tf.keras.layers.Conv2D(64, (3, 3), activation='relu'),   # 64 filters,
    #   3x3 kernel
    tf.keras.layers.Flatten(),   # Flatten the 3D output to 1D
    tf.keras.layers.Dense(64, activation='relu'),
    # Fully connected layer with 64 units
    tf.keras.layers.Dense(10, activation='softmax')   # Output layer with
    #   10 units (one for each digit)
])

# Compile the model
model.compile(optimizer='adam',
```

5.2 Convolutional Neural Networks

```
              loss='sparse_categorical_crossentropy',
              metrics=['accuracy'])

# Train the model
model.fit(x_train, y_train, epochs=5)

# Evaluate the model on the test set
test_loss, test_acc = model.evaluate(x_test, y_test)
print(f'\nTest accuracy: {test_acc:.4f}')

# Use the model to make predictions on the test set
predictions = model.predict(x_test)

# Example: Print the predicted label for the first test image
print(f'Predicted label for the first test image:
    {predictions[0].argmax()}')
```

The code loads the MNIST dataset using TensorFlow's Keras API. The images are normalized to have pixel values between 0 and 1 and reshaped to include a single channel (28x28x1) since CNNs expect a 3D input in vector form (height, width, channels). The CNN architecture involves first a `Conv2D Layer` that applies convolutional filters to the input. The first layer has 32 filters with a 3x3 kernel, followed by ReLU activation function. Next a `MaxPooling2D Layer` reduces the input spatial dimensions using a 2x2 filter size. Another `Conv2D Layer` follows with 64 filters and a 3x3 kernel. Next, `MaxPooling2D Layer`, i.e. another max pooling layer followed by `Conv2D Layer`, i.e. a third convolutional layer with 64 filters and a 3x3 kernel. The `Flatten Layer` flattens the 3D output from the last convolutional layer into a 1D vector. The `Dense Layer` that follows is a fully connected layer with 64 neurons and ReLU activation and last the `Output Dense Layer` is the final layer has 10 neurons, one for each digit, with softmax activation to produce probabilities.

The model is compiled using the Adam optimizer and the sparse categorical crossentropy loss function, suitable for multi-class classification. It is trained on the training data for 5 epochs and its accuracy is evaluated on the test set. The model is used to make predictions on the test set, and the predicted label for the first test image is printed. After running the model we obtain test set accuracy equal to 0.9908

The MNIST classification can be performed with simple feed forward neural networks as well but the CNN architecture is more powerful than a simple neural network and achieves better accuracy on image classification tasks.

5.3 Recurrent Neural Networks

A recurrent neural network or RNN has architecture that enables it to process sequential data such as time series, language modeling, text generation, etc although it can handle many other tasks such as classification as well. Unlike feed forward networks or CNNs the RNNs have connections that cycle back output and thus retain some form of memory. In Fig. 5.5 we see an example of an RNN where in addition to the input at time t there is also the output at the previous time step $t - 1$ that is taken into account.

The RNNs have a number of features that differentiates them from the networks we considered so far. They are designed to handle sequential data while maintaining a *hidden state* that carries information from previous inputs in the sequence. At each time step in a sequence, an RNN has a hidden state that is updated based on the current input and the previous value of the hidden state. This hidden state acts as memory cell that carries information forward through the sequence. The presence of the hidden state makes RNNs *stateful* and not stateless as other networks considered so far. All RNNs have a thus a feedback mechanism that makes them more complex than other networks. They typically use sigmoid or tanh activation function. A special type of RNNs that include multiple gates is the Long Short Term Memory networks or LSTMs. The LSTMs are designed to better capture long-term dependencies by using a more complex structure, including gates that control the flow of information. LSTMs are effective at handling the vanishing gradient problem that appears during training in RNNs. The training of RNNs involves a variation of the backpropagation algorithm called Backpropagation Through Time (BPTT). In BPTT, gradients are calculated and propagated back through all the previous time steps; this enables the model to learn from the entire sequence. The BPTT can suffer from problems like the vanishing gradient problem, where gradients become very small, making it difficult for the network to learn long-term dependencies.

The central mathematical formula that describes RNNs is the equation of the hidden memory at time t, i.e. h_t:

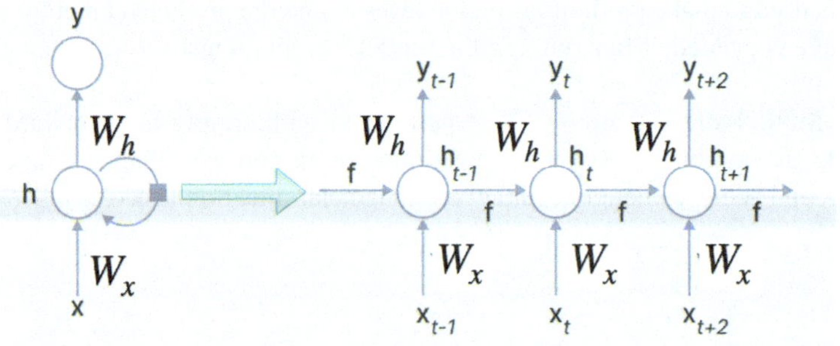

Fig. 5.5 Example of an RNN unfolded in time

5.3 Recurrent Neural Networks

$$h_t = f(W_h h_{t-1} + W_x x_t + b) \tag{5.3}$$

where h_{t-1} is the hidden state from the previous time step, x_t is the input at the current time step, W_h and W_x are weight matrices, b is the bias vector and f is the activation function such as tanh or ReLU. The output at each time step is computed based on the current hidden state.

In the code below we show simple example of the usage of RNNs through TensorFlow and Keras. The network is trained on a sequence of integers and then it is asked to predict the next digit in a different sequence.

```python
import tensorflow as tf
import numpy as np

# Create a simple dataset: a sequence of numbers
# We'll create sequences like [0, 1, 2, 3, 4] -> [5], [1, 2, 3, 4, 5] ->
# [6], etc.
sequence_length = 5
dataset_size = 100

x_train = []
y_train = []

for i in range(dataset_size):
    x_train.append(np.arange(i, i + sequence_length))
    y_train.append(i + sequence_length)

x_train = np.array(x_train)
y_train = np.array(y_train)

# Reshape input to be [samples, time steps, features]
x_train = x_train.reshape((x_train.shape[0], x_train.shape[1], 1))

# Build a simple RNN model
model = tf.keras.models.Sequential([
    tf.keras.layers.SimpleRNN(50, activation='relu',
        input_shape=(sequence_length, 1)), tf.keras.layers.Dense(1)])

# Compile the model
model.compile(optimizer='adam', loss='mse')

# Train the model
model.fit(x_train, y_train, epochs=200, verbose=0)

# Test the model: predict the next number in a new sequence
new_sequence = np.array([100, 101, 102, 103, 104]).reshape((1,
    sequence_length, 1))
predicted_value = model.predict(new_sequence)
print(f'Predicted next value in sequence [100, 101, 102, 103, 104]:
    {predicted_value[0][0]:.2f}')
```

In the code the input dataset consists of sequences of numbers. For example, [0, 1, 2, 3, 4] predicts 5, [1, 2, 3, 4, 5] predicts 6, etc. The x_{train} contains the input sequences, and y_{train} contains the corresponding next values in the sequence. The RNNs expect input data in the form (samples, time steps, features). Here, each sequence is a sample, the length of the sequence is the number of time steps, and there is one feature per time step, i.e. the number itself.

One RNN layer with 50 units, i.e. fifty memory cells or neurons is used that processes the input sequence. A fully connected dense layer follows with one output unit that aims at predicts the next value in the sequence. The model uses the Adam optimizer and Mean Squared Error (MSE) as the loss function that are suitable for regression tasks like this one. It is trained on the synthetic sequence data for 200 epochs. After training, the model is tested with a new sequence [100, 101, 102, 103, 104] to predict the next number in the sequence. After running the code we obtain the result *Predicted next value in sequence [100, 101, 102, 103, 104]: 105.30*. This simple RNN is able to come close to predicting the next value in an unknown sequence.

5.4 Long Short-Term Memory Networks

The Long Short-Term Memory or LSTM networks are a type of RNN that is well suited for NLP, speech recognition and handwriting recognition [3]. They can learn *long term dependencies*, i.e. the distance gap between relevant information and the location where it is needed. In these cases the information in one part of a document must be linked to another distant location in the document. The LSTMs are *stateful* and contain three gates, the *forget gate*, the *input gate* and an *output gate* that use the sigmoid activation function as well as a *cell state* that involves the tanh activation function (Fig. 5.6).

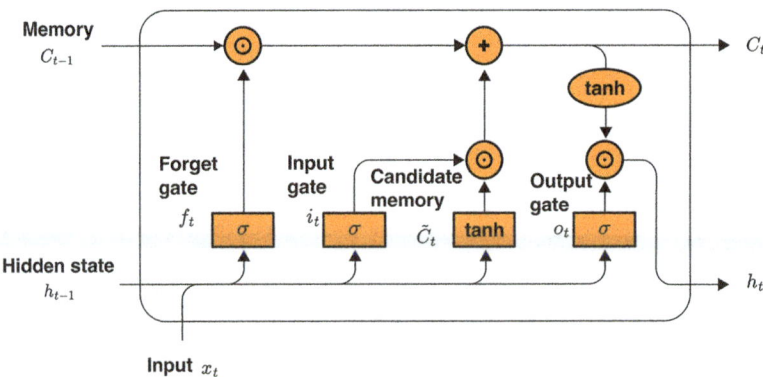

Fig. 5.6 An LSTM cell

5.4 Long Short-Term Memory Networks

The central idea behind LSTMs is the memory cell that can maintain its state over long periods of time. The memory cell has the ability to decide what information to keep, what to forget, and what to output at each time step. The LSTMs have a special structure made up of three gates, which control the flow of information. The forget Gate decides what information to discard from the cell state. It takes the previous hidden state and the current input and passes them through a sigmoid function to produce a number between 0 and 1 for each number in the cell state. A value close to 0 means "forget this" and a value close to 1 means "keep this." The functional form is

$$f_t = \sigma(W_f \cdot [h_{t-1}, x_t] + b_f) \tag{5.4}$$

The Input Gate decides which new information to add to the cell state. It consists of two parts: the input gate layer, which uses a sigmoid function, and a tanh layer that creates a vector of new candidate values. These two parts are combined to update the cell state.

$$i_t = \sigma(W_i \cdot [h_{t-1}, x_t] + b_i) \tag{5.5}$$

$$\tilde{C}_t = \tanh(W_C \cdot [h_{t-1}, x_t] + b_C) \tag{5.6}$$

The output gate decides what the next hidden state should be. The hidden state is also used for predictions. The output gate controls which parts of the cell state are sent in the output.

$$o_t = \sigma(W_o \cdot [h_{t-1}, x_t] + b_o) \tag{5.7}$$

The cell state acts as a highway that runs through the entire chain, with only some minor linear interactions. It carries information across the entire sequence, allowing the network to remember information over long periods. The cell state is updated using the forget gate and input gate.

$$C_t = f_t \cdot C_{t-1} + i_t \cdot \tilde{C}_t \tag{5.8}$$

The hidden state is the output of the LSTM at each time step. It is based on the current cell state and the output gate. The hidden state is passed on to the next LSTM cell and can be used for making predictions.

In the LSTM formulas of Equations (5.4) to (5.8) we denote with σ the sigmoid activation function and tanh the hyperbolic tangent activation function. The weight matrices W_f, W_i, W_C, W_o are for the forget gate, input gate, candidate cell state, and output gate, respectively while b_f, b_i, b_C, b_o are the corresponding bias vectors for each respective gate. The term h_{t-1} denotes the previous hidden state, x_t is the current input while C_t is the cell state at time step t. The LSTM formulas represent the core computations within an LSTM cell. The operation of each gate helps control the flow of information through the network and permits LSTMs to effectively learn from sequences with long-term dependencies.

As an example we use LSTM in order to predict the next digit in a sequence; the code is similar to the previous one with general RNNs:

```python
import numpy as np
import tensorflow as tf

# Create a simple dataset: sequences of numbers
# We'll create sequences like [0, 1, 2, 3, 4] -> [5], [1, 2, 3, 4, 5] ->
# [6], etc.
sequence_length = 5
dataset_size = 100

x_train = []
y_train = []

for i in range(dataset_size):
    x_train.append(np.arange(i, i + sequence_length))
    y_train.append(i + sequence_length)

x_train = np.array(x_train)
y_train = np.array(y_train)

# Reshape input to be [samples, time steps, features]
x_train = x_train.reshape((x_train.shape[0], x_train.shape[1], 1))

# Build the LSTM model
model = tf.keras.models.Sequential([
    tf.keras.layers.LSTM(50, activation='relu',
        input_shape=(sequence_length, 1)), tf.keras.layers.Dense(1)
])

# Compile the model
model.compile(optimizer='adam', loss='mse')

# Train the model
model.fit(x_train, y_train, epochs=200, verbose=0)

# Test the model: predict the next number in a new sequence
new_sequence = np.array([100, 101, 102, 103, 104]).reshape((1,
 sequence_length, 1)) predicted_value = model.predict(new_sequence)
print(f'Predicted next value in sequence [100, 101, 102, 103, 104]:
    {predicted_value[0][0]:.2f}')
```

The LSTM layer has 50 units and processes the input sequence using the ReLU activation function. The aim, as before is to train the network to predict the next value in an unknown sequence of numbers. After running the code we find *Predicted next value in sequence [100, 101, 102, 103, 104]: 105.00*. This is the correct result and the answer is superior to the one given by an RNN previously.

5.5 Conclusions

The CNNs use both convolutions and forms of course-graining to process effectively image classification and object detection. They are designed to learn spatial hierarchies of features such as edges, textures, and shapes. They are used in image classification, object detection, and segmentation and are quite efficient due to parameter sharing in convolutional layers. The RNNs, unlike other networks, use cycles in connections to remember previous inputs over time. They are designed for processing sequential data such as time-series or text. Applications include language modeling, speech recognition, and machine translation. They are also widely used in various scientific domains to model sequential data and time-dependent processes. In physics and climate science, RNNs are employed to predict chaotic systems, simulate weather patterns, and forecast environmental changes by capturing temporal dependencies in time-series data. In bioinformatics, RNNs are utilized for analyzing sequential biological data, such as DNA, RNA, and protein sequences, where recognizing dependencies in genetic sequences is critical for understanding biological functions and mutations. In neuroscience, RNNs help model neural activity over time and are used to analyze brain signal data, such as electroencephalograms (EEG), to detect patterns in neurological conditions. RNNs are also applied in finance to forecast stock prices and economic trends by modeling temporal relationships in market data. Despite their utility, RNNs can struggle with capturing long-term dependencies in complex scientific data, which is why advanced architectures like LSTMs and GRUs are often preferred in many scientific applications.

Finally, LSTM networks use special gates to control the flow of information, retaining long-term dependencies. They are useful in tasks requiring long-term memory, such as time-series forecasting, machine translation, and text generation while they are also popular in natural language processing and speech recognition. In science, LSTM networks are employed for long-term weather forecasting, they are effective for analyzing biological sequences, such as DNA, RNA, and protein structures. They are applied to model brain activity patterns and to analyze time-series data from electroencephalogram and functional magnetic resonance imaging (fMRI) for detecting neurological disorders. In physics are employed in complex dynamical systems and predicting chaotic processes.

5.6 Summary

- CNNs are designed for processing grid-like data such as images. Are commonly used in image classification, object detection, and segmentation.
- RNNs are designed for processing sequential data like time-series or text. They use cycles in connections to remember previous inputs over time.

- RNNs common applications include language modeling, speech recognition, and machine translation.
- LSTM network is a special type of RNN designed to overcome the vanishing gradient problem. Utilize gates to control the flow of information, retaining long-term dependencies.
- LSTMs are effective in tasks requiring long-term memory, such as time-series forecasting, machine translation, and text generation.

References

1. A. Géron, *Hands-on Machine Learning with Scikit-Learn, Keras and TensorFlow*, 3rd edn. (O'Reilly Media, Inc., 2023)
2. I. Goodfellow, Y. Bengio, A. Courville, *Deep Learning* (MIT Press, 2016)
3. S. Hochreiter, J. Schmidhuber, Long short-term memory. Neural Comput. **9**(8), 1735–1780 (1997)

Chapter 6
Autoencoders and More

Autoencoders, Reservoir Computing, and PINNs

Abstract In this chapter we present more specialized but very useful techniques. Autoencoders perform information processing though a bottleneck that determines the true information content of the data. They are simple mirror image-type network designs. Reservoir computing uses a random network reservoir that, while not trained, it provides useful states that help the general learning process. Physics informed networks can combine data with dynamical equations and learn from both.

6.1 Introduction

Autoencoders are a type of neural network designed to learn efficient representations of data, typically for the purpose of dimensionality reduction, feature extraction, or denoising. In scientific and industrial applications, autoencoders are used for tasks such as image compression, where the network learns to encode high-dimensional images into a smaller latent space and reconstruct them with minimal loss. In anomaly detection, autoencoders are trained to reconstruct normal data patterns, and anomalies are identified when the reconstruction error exceeds a certain threshold. Autoencoders are also used in the pretraining of neural networks, particularly for unsupervised learning, by learning a compact representation of the input data before fine-tuning with labeled data. Variants of autoencoders, such as denoising autoencoders, are utilized to remove noise from data, improving signal quality in fields like image processing and speech recognition. In general, autoencoders are a powerful tool for learning compressed, meaningful representations of data in a variety of domains.

Reservoir computing is a framework for efficiently processing temporal data and capturing complex dynamic behavior in systems. It leverages a large, fixed recurrent neural network called the reservoir, which projects input data into a high-dimensional space. Only the output layer is trained, allowing for quick and efficient learning. This technique is particularly useful for real-time tasks and chaotic time-series prediction, and it has been applied to a variety of scientific fields, including fluid dynamics, biological signal processing, and speech recognition. Its lightweight training process and ability to handle nonlinear systems make reservoir computing attractive for applications that require fast, adaptive learning in complex environments.

Physics informed machine learning (PIML) integrates physical laws and constraints into machine learning models, allowing them to respect the underlying principles of scientific systems, such as conservation laws or partial differential equations (PDEs). This hybrid approach enhances data-driven models by incorporating prior knowledge of the system's behavior, improving their accuracy and generalizability, especially in scenarios with limited data. PIML is particularly valuable in fields like fluid dynamics, climate modeling, and structural analysis, where it enables researchers to create models that are not only more accurate but also more interpretable. By embedding physics into neural networks, PIML reduces the risk of overfitting and produces models that remain consistent with known physical theories.

6.2 Autoencoders

Autoencoders are neural networks that are used for unsupervised learning, particularly for tasks like data compression, noise reduction, and feature learning [1, 2]. They consist of two parts, the *encoder* that learns a representation of the input data and the *decoder* that reproduces it in the output. In between encoder and decoder there is a single hidden layer that has fewer neurons than either the input or the output layer. This is the *latent space* that effects data compression resulting in dimensionality reduction. In many ways the autoencoder acts as a unit operator to data, while they pass through the information bottleneck of the latent space (Fig. 6.1).

The encoder compresses the input into a lower-dimensional representation. It maps the input data x to the latent space h. This part of the network is typically a series of layers that reduce the dimensionality of the input, capturing the essential features. The mathematical representation of the encoder is

$$h = f(x), \qquad (6.1)$$

where h is the encoded representation and f is the function learned by the network. The latent space that follows the encoder provides a compressed, lower-dimensional representation of the input data that captures in principle the most important features of the data. Finally, the decoder reconstructs the input data from the encoded representation. It maps the latent space h back to the original input space $\hat{x}$. The goal is to make the reconstruction $\hat{x}$ as close to the original input x as possible. The decoder is represented by the functional form

$$\hat{x} = g(h), \qquad (6.2)$$

6.2 Autoencoders

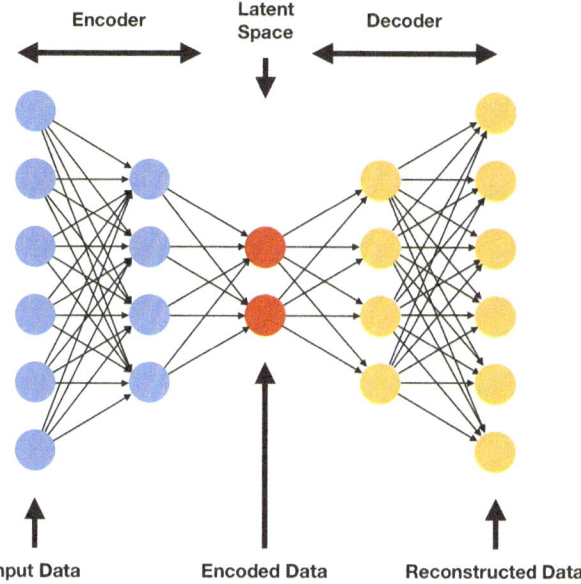

Fig. 6.1 Autoencoder

where g is the function that is learned by the neural network. During training the autoencoder aims at minimizing the distance between input and x and reconstructed output $\hat{x}$ through a loss function such as the mean squared error one:

$$Loss = ||x - \hat{x}||^2. \tag{6.3}$$

6.2.1 Autoencoders and PCA

The autoencoders are used for classification, document retrieval, anomaly detection, image denoising, etc. They can also be used in order to extract information similar to that of the principal component analysis. This is possible when the autoencoder involves linear activations or only a single sigmoid hidden layer. In the code that follows we use an autoencoder to perform PCA.

```
import numpy as np
import tensorflow as tf
from tensorflow.keras.layers import Input, Dense
from tensorflow.keras.models import Model
from sklearn.datasets import load_iris
from sklearn.preprocessing import StandardScaler
import matplotlib.pyplot as plt
```

```python
# Load the dataset (e.g., Iris dataset)
data = load_iris()
X = data.data

# Standardize the data (important for PCA and autoencoders)
scaler = StandardScaler()
X_scaled = scaler.fit_transform(X)

# Define the dimensions
input_dim = X_scaled.shape[1]  # Number of features
encoding_dim = 2  # Dimension for PCA (reduce to 2 components)

# Build the autoencoder model
input_layer = Input(shape=(input_dim,))
encoded = Dense(encoding_dim, activation='linear')(input_layer)
# Encoding layer (linear activation for PCA)
decoded = Dense(input_dim, activation='linear')(encoded)
# Decoding layer

# Model that maps input to reconstruction
autoencoder = Model(inputs=input_layer, outputs=decoded)

# Model that maps input to encoded representation
encoder = Model(inputs=input_layer, outputs=encoded)

# Compile the model
autoencoder.compile(optimizer='adam', loss='mse')

# Train the autoencoder
autoencoder.fit(X_scaled, X_scaled, epochs=100, batch_size=16,
    shuffle=True, verbose=0)

# Use the encoder to get the compressed representation
X_encoded = encoder.predict(X_scaled)

# Plot the encoded data (equivalent to PCA)
plt.figure(figsize=(8, 6))
plt.scatter(X_encoded[:, 0], X_encoded[:, 1], c=data.target,
    cmap='viridis')
plt.colorbar()
plt.title('Autoencoder Output (2D Encoding, Similar to PCA)')
plt.xlabel('Encoded Feature 1')
plt.ylabel('Encoded Feature 2')
plt.show()
```

In this code we use the Iris dataset and standardize it using StandardScaler. Standardization is important for both PCA and autoencoders because it ensures that each feature contributes equally to the analysis. The autoencoder has an input layer matching the number of features in the dataset. The encoding layer reduces the data to a lower-dimensional space that in this case is two dimensional—this is analogous to the principal components in PCA. The decoding layer attempts to reconstruct the

Fig. 6.2 Autoencoder PCA: The encoded data are visualized using a scatter plot. The 2D encoding learned by the autoencoder should capture the key structure of the data, similar to how the first two principal components would be in PCA

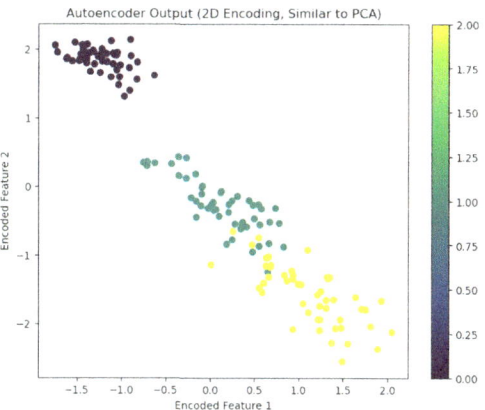

original input from the encoded representation. During training the autoencoder tries to minimize the reconstruction loss while forcing the encoder to learn an efficient, lower-dimensional representation of the input data. After training the encoder part of the model compresses the data representation into PCA-like components. The separation of the classes (Fig. 6.2) is similar to the one shown in Fig. 3.1 with PCA in Chap. 3.

6.2.2 Variational Autoencoders

Variational autoencoders (VAEs) combine the principles of autoencoders with probabilistic modeling. Unlike traditional autoencoders, which map inputs to a fixed latent space, VAEs map inputs to a distribution in the latent space, allowing for the generation of new data by sampling from this distribution. While the encoder and decoder have similar structure as in the standard autoencoders in the variational ones, both sides have a probability distribution associated with the encoding and decoding processes. The latent space in VAE is also probabilistic since the encoder maps the input data to parameters of a probability distribution—typically a Gaussian distribution—in the latent space. Instead of encoding an input x to a single point z in the latent space, it encodes x to a mean μ and a standard deviation σ of a Gaussian distribution in the latent space. During training, the latent vector z is sampled from the Gaussian distribution parameterized by μ and σ. This sampling introduces stochasticity into the model, allowing it to generate diverse outputs from the same input distribution. There are various technical issues related to VAEs but will not be discussed further here. The VAEs have many applications including in data and speech generation, anomaly detection, dimensionality reduction, Bayesian inference, etc.

6.3 Reservoir Computing

The reservoir computing is a computational framework that uses the dynamics of a fixed, complex system termed "reservoir" in order to process input data [3]. It is used for time-series data and other sequential data. Reservoir computing simplifies the training process by only requiring the training of a readout layer, while the reservoir itself remains untrained and fixed. The reservoir is a large, recurrent neural network with fixed random connections and maps the input sequence into a high-dimensional space. The key idea is that the reservoir's internal state changes in response to the input and thus is able to capture the temporal dynamics of the data. The reservoir can be thought of as a "liquid" that reacts to perturbations, i.e., the inputs and whose internal states can be thought as ripples in the liquid and contain information about the inputs over time.

The input data are fed into the reservoir through an input layer that connects the inputs to the nodes in the reservoir. The connections between the input and the reservoir are fixed and initialized randomly. As the input data passes through the reservoir, the nodes in the reservoir network update their states based on the input and the states of neighboring nodes. The reservoir's rich internal dynamics allow it to capture complex temporal patterns and dependencies in the input data. The readout layer is a simple, typically linear, layer that reads the states of the nodes in the reservoir and produces the final output. This layer is the only part of the system that is trained. In Fig. 6.3 we present a graph with the workings of reservoir computing.

There are two popular forms of reservoir computing. In the Echo State Network (ESN) the reservoir is a recurrent neural network with random, fixed connections. In ESNs the influence of past inputs diminishes over time. The Liquid State Machines (LSMs) are inspired by biological neural networks. The reservoir in LSMs consists of spiking neurons, and the dynamics is based on the timing of the spikes. The main advantage of reservoir computing is that only the readout layer is trained, a fact that reduces significantly the computational complexity compared to traditional recurrent neural networks (RNNs) where all weights are trained. This makes reservoir computing particularly attractive for tasks that require processing

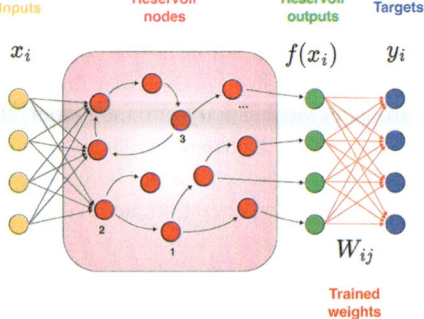

Fig. 6.3 Reservoir computing includes a bath of neurons designated as 1, 2, 3, . . . that are not trained

6.3 Reservoir Computing

large amounts of sequential data. Reservoir computing is used in a wide range of applications, from time-series prediction and signal processing to control systems and cognitive computing.

The following code is based on reservoir computing:

```python
import numpy as np
import matplotlib.pyplot as plt
from pyESN import ESN

# Generate a simple time-series (sine wave)
def generate_sine_wave(length, frequency=0.1):
    return np.sin(2 * np.pi * frequency * np.arange(length))

# Generate training data
train_length = 200
train_data = generate_sine_wave(train_length)

# Generate testing data
test_length = 100
test_data = generate_sine_wave(test_length, frequency=0.1)

# Initialize the Echo State Network (ESN)
esn = ESN(n_inputs=1, n_outputs=1, n_reservoir=500, spectral_radius=1.25,
    sparsity=0.2, random_state=42)

# Train the ESN on the training data
esn.fit(train_data[:-1].reshape(-1, 1), train_data[1:].reshape(-1, 1))

# Use the ESN to predict the next values in the sequence
predicted = esn.predict(test_data[:-1].reshape(-1, 1))

# Plot the results
plt.figure(figsize=(10, 5))
plt.plot(range(train_length, train_length + test_length - 1),
    test_data[1:], label='True Signal')
plt.plot(range(train_length, train_length + test_length - 1),
    predicted, label='Predicted Signal')
plt.title("Time-Series Prediction using Echo State Network")
plt.xlabel("Time Step")
plt.ylabel("Signal Value")
plt.legend()
plt.show()
```

In order to run this code we need first to install pyESN, i.e., use pip install pyESN in the command line. The code generates a simple sine wave and used this as input data. We then create an ESN with one dimensional input and output, spectral radius, a parameter that contains reservoir dynamics equal to 1.25 and sparsity of connections to 20% level. The fit method is used to train the ESN on the training data. The model learns to predict the next value in the time series based on the

Fig. 6.4 Reservoir computing results

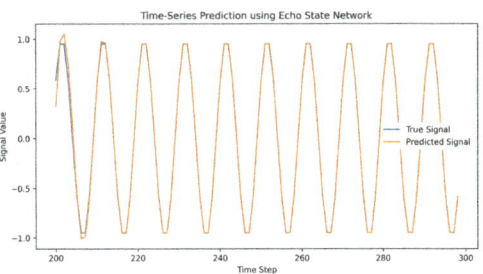

current value. When we run this code, it generates a sine wave, train an Echo State Network on the initial portion of the wave, and then predict the next values in the sequence. The results are plotted in Fig. 6.4.

6.4 Physics Informed Machine Learning

The PIML combines machine learning techniques with physical laws and principles. The goal is to create models that not only learn from data but also respect the underlying physics governing the system. This approach is particularly useful in cases where standard machine learning models have difficulties due to limited data or the need for physical interpretability.

In PIML, physical laws, such as conservation of energy, momentum, or mass, are embedded into the learning process. This can be done by encoding these laws as constraints in the machine learning model. For example, the loss function in a neural network might include terms that penalize deviations from known physical laws, such as the Navier-Stokes equations in fluid dynamics. PIML often involves hybrid models that combine data-driven components like neural networks with traditional physics-based models that are usually modeled through differential equations. These hybrid models can take advantage of the strengths of both approaches, i.e., the ability of machine learning to learn from complex data and the reliability of physics-based models.

The Physics Informed Neural Networks (PINNs) are a popular framework within PIML where neural networks are trained to solve PDEs by incorporating them directly into the loss function [4, 5]. In this approach, the neural network learns to approximate the solution to a PDE while respecting boundary conditions and other physical constraints.

When we use PINNs the neural network is trained to satisfy both the data and the physical laws described by differential equations. Let us consider the case where we have data that stem from particles in one dimension executing diffusive motion. The PDE equation describing one dimensional diffusion equation is

$$\frac{\partial u}{\partial t} = D \frac{\partial^2 u}{\partial x^2}, \qquad (6.4)$$

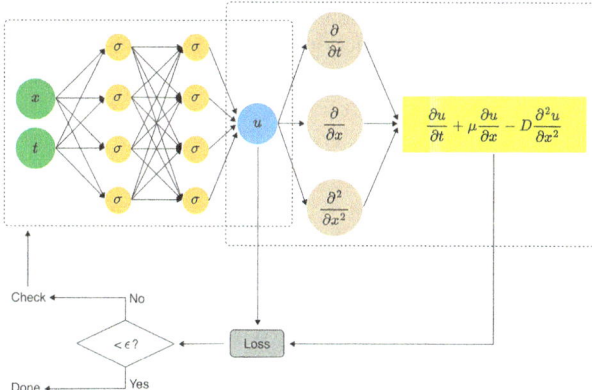

Fig. 6.5 Physics Informed Machine Learning uses a neural network, and in the loss function one adds a term that comes from the differential equation [5]. In the figure for $\mu = 0$, the loss term checks that the data fit the diffusion equation $u_t = D u_{xx}$

where $u(x, t)$ is the density of particles in position x at time t and D is the diffusion constant. In order to train a neural network to the available data while also respecting the differential equation, we may augment the loss function as follows:

$$Loss = MSE(data) + MSE(\frac{\partial u}{\partial t} - D\frac{\partial^2 u}{\partial x^2}). \tag{6.5}$$

This form of the loss function ensures that the network learns to produce solutions that are consistent with the physical law represented by the PDE, i.e., the diffusion equation. The process is shown in Fig. 6.5.

Physics informed machine learning is a powerful approach that merges the strengths of machine learning with the rigor of physical laws. By embedding physics into the learning process, PIML models can achieve higher accuracy and better generalization. The application of PIML in complex systems may provide a useful approach especially when the amount of data available is small.

6.5 Conclusion

Autoencoders are a type of artificial neural network used primarily for unsupervised learning, where the goal is to encode input data into a lower-dimensional representation and then decode it to reconstruct the original data. They are widely applied in scientific fields for tasks such as dimensionality reduction, anomaly detection, and data denoising. In bioinformatics, autoencoders are used for compressing high-dimensional genetic data, enabling researchers to analyze large-scale genomic datasets more efficiently. In physics, autoencoders are used for detecting anomalies in particle physics experiments, where they can identify rare events by learning normal data distributions and flagging deviations. Autoencoders are also useful for image reconstruction and noise reduction in medical imaging, where they improve the quality of MRI or CT scans by removing noise. Additionally, they are employed in environmental science for feature extraction from complex climate datasets, enabling better climate prediction models. Their ability to learn compressed, informative representations makes autoencoders valuable in a range of scientific applications.

Reservoir computing is a framework for efficiently processing temporal data and capturing complex dynamic behavior in systems. It leverages a large, fixed recurrent neural network called the reservoir, which projects input data into a high-dimensional space. Only the output layer is trained, allowing for quick and efficient learning. This technique is particularly useful for real-time tasks and chaotic time-series prediction, and it has been applied to a variety of scientific fields, including fluid dynamics, biological signal processing, and speech recognition. Its lightweight training process and ability to handle nonlinear systems make reservoir computing attractive for applications that require fast, adaptive learning in complex environments.

PIML integrates physical laws and constraints into machine learning models, allowing them to respect the underlying principles of scientific systems, such as conservation laws or PDEs. This hybrid approach enhances data-driven models by incorporating prior knowledge of the system's behavior, improving their accuracy and generalizability, especially in scenarios with limited data. PIML is particularly valuable in fields like fluid dynamics, climate modeling, and structural analysis, where it enables researchers to create models that are not only more accurate but also more interpretable. By embedding physics into neural networks, PIML reduces the risk of overfitting and produces models that remain consistent with known physical theories.

6.6 Summary

- Autoencoders reconstruct the input while reducing the nodes necessary to a minimum in the latent space.
- Reservoir computing uses a random neural network to process information.
- Physics informed machine learning uses data and differential equations to train neural networks.

References

1. A. Géron, *Hands-on Machine Learning with Scikit-Learn, Keras and TensorFlow*, 3rd edn. (O'Reilly Media, Inc., 2023).
2. I. Goodfellow, Y. Bengio, A. Courville, *Deep Learning* (MIT Press, 2016)

References

3. H. Jaeger, H. Haas. Harnessing nonlinearity: predicting chaotic systems and saving energy in wireless telecommunication. Science **308**, 78–80 (2004)
4. I.E. Lagaris, A. Likas, D.I. Fotiadis, Artificial neural networks for solving ordinary and partial differential equations. IEEE Trans. Neural Networks **9**, 5 (1998)
5. M. Raissi, P. Perdikaris, G.E. Karniadakis, Physics-informed neural networks: A deep learning framework for solving forward and inverse problems involving nonlinear partial differential equations. J. Comput. Phys. **378**, 686–707 (2019)

Part III
Discrete Nonlinear Models

Chapter 7
The Discrete Nonlinear Schrödinger Equation

From Biomolecules to Nonlinear Optics to Bose-Einstein Condensates

Abstract The discrete nonlinear Schrödinger equation is a ubiquitous and fundamental equation in nonlinear physics and mathematics. It describes properties of chemical, condensed matter, optical, and other systems where self-trapping mechanisms are present. The latter arise from either strong interaction with the environment or genuine nonlinear properties of the medium. The continuous nonlinear Schrödinger equation is one of the few nonlinear partial differential equations that are analytically solvable. The discrete version we are discussing here cannot be solved analytically in general except for some special cases. One of these is the nonlinear dimer, a system that has only two sites and where an excitation may tunnel from one to the next. The analytical solution is expressed through elliptic functions, the Jacobian functions, or the Weierstrass elliptic function. In the case where both nonlinear sites have the same energy and the excitation is localized initially on one site, we find that there is a self-trapping transition for a specific value of the nonlinearity parameter. For this value the elliptic function time evolution becomes hyperbolic, signaling a true change in the excitation dynamics.

7.1 The DNLS Equation

The discrete nonlinear Schrödinger or the DNLS equation was introduced by Eilbeck, Lomdahl, and Scott as the *Discrete Self-Trapping* equation [1] in the context of the *Davydov soliton* problem [2, 3]. Davydov discussed in the 1970s and 1980s the idea that excitations propagating in biological macromolecules such as proteins must interact strongly with vibrational modes, and, as a result, a new coupled entity is formed, that of a *soliton*. This scenario would be very beneficial for energy transport in complex media such as proteins since solitons transport energy without dispersion. If this is the case, then an initially deposited energy packet in part of the system may move without distortion in some other part of the molecule at relatively large distances and be used there. Proteins are molecular systems with units that interact weakly among them, and thus it makes more sense to describe these dynamic phenomena in a weakly coupling limit; this leads to discrete equations rather than continuous ones. The DNLS equation is a discrete version of the continuous equation that has soliton solutions. It includes hopping in nearby discrete sites, while it also incorporates nonlinearity; this combination of terms leads to a self-trapping mechanism analogous to the soliton in the continuous systems. The mathematical form we use for the DNLS equation is the following:[1]

$$i\frac{d\psi_n}{dt} = \epsilon_n \psi_n + V(\psi_{n-1} + \psi_{n+1}) - \chi_n |\psi_n|^2 \psi_n. \tag{7.1}$$

The basic quantity of interest is $\psi_n(t)$ that denotes the probability amplitude for a quantum mechanical particle, such as an electron or an *exciton*, i.e., a local molecular excitation, to be found at the site n at time t. The physical picture we have is the following: The exciton behaves like a quantum mechanical particle that propagates in a wave-like way following the basic equation of microscopic physics, i.e., the Schrödinger equation. This exciton then hops from molecule to molecule and "lands" at specific energy levels of each molecule. For simplicity we take the propagation to be in a one dimensional lattice and assume that each molecule has one energy level available to the exciton; the local energy at the n-th molecule or site is ϵ_n, while V is the coupling from each molecule to its neighbor that is the term that enables transfer. This form of the equation without the nonlinear term is fully quantum mechanical and is usually referred to as the tight binding approximation to the Schrödinger equation [4]. Its quantum nature is manifested also from the presence of $\hbar$ that should be on the LHS of the equation as $i\hbar \frac{d\psi_n}{dt}$. For reasons of simplicity we take $\hbar = 1$.

The term that makes the DNLS very interesting is the nonlinear term $-\chi_n |\psi_n|^2 \psi_n$. We note that this term is in some sense similar to the first term on the RHS of the equation; indeed if we write $\epsilon'_n = -\chi_n |\psi_n|^2$, we may have the first and third terms of the equation formally identical. But there is a major difference, while in the first term the local energy level ϵ_n is *fixed*, and in the third, nonlinear, term the local energy ϵ'_n depends on $|\psi_n|^2$. The latter is nothing but the probability to find the particle at the molecular site n. Thus ϵ'_n is *variable*, and its actual value depends on how high is the probability to have the particle in that site. If this probability is high enough, then self-trapping results.

The source of this very important nonlinear term is the coupling of the electronic excitation with vibrational modes in the molecular system. The exact equation in this case is much more complicated, but the use of a *diabatic approximation* leads to the simpler but very powerful DNLS equation [5].

---

[1] This is a slightly more general form than the DNLS equation presented in Chap. 1. In Eq. (1.6) all nonlinearity terms are identical and equal to χ.

7.2 Diabatic Approximation

For the derivation of the DNLS equation in condensed matter context start with the Hamiltonian:

$$H = \sum_n \epsilon_n |n\rangle\langle n| - J \sum_n [|n+1\rangle\langle n| + |n\rangle\langle n+1|]$$
$$+ \frac{K}{2} \sum_n u_n^2 + \frac{1}{2} M \sum_n (du_n/dt)^2 - \sum_n A_n u_n |n\rangle\langle n| \quad (7.2)$$

representing an excitation moving in a one-dimensional crystal while interacting with local, Einstein-type, oscillators. In Eq. (7.2) ϵ_n represents the local site energy at site n, J gives the magnitude of the wave function overlaps of neighboring sites, $|n\rangle$ and $\langle n|$ are related to the probability amplitudes at site n, whereas u_n is the displacement of the n-th local oscillator. The exciton-phonon coupling term is diagonal in the $|n>$ basis and depends only on local oscillator displacements. We assume that the exciton-phonon coupling A_n is different for each molecular site n.

We may now neglect the kinetic energy vibrational terms and expand the time dependent wave function as $|\psi\rangle = \sum_p \psi_p |p\rangle$, where the $|p\rangle$ represent Wannier states. Inserting this into the time dependent Schrödinger equation $i(d|\psi\rangle/dt) = H|\psi\rangle$ and using the orthonormality property for the $|p\rangle$, we obtain

$$i\frac{d\psi_n}{dt} = \epsilon_n \psi_n - J[\psi_{n-1} + \psi_{n+1}] - A_n u_n \psi_n + \frac{K}{2} \sum_m u_m^2 \psi_n.$$

In the next step we eliminate the vibrational degrees of freedom by imposing the condition of minimization of the energy of the stationary states. Inserting the time dependent part of $\psi_n \sim \exp[iEt]$ and using the normalization condition for the amplitudes ψ_p, $\sum_p |\psi_p|^2 = 1$, we obtain for the total excitation and lattice energy E:

$$E = \sum_n [\epsilon_n - A_n u_n] |\psi_n|^2 - J \sum_n (\psi_{n-1} - \psi_{n+1}) \psi_n^* + \frac{K}{2} \sum_n u_n^2. \quad (7.3)$$

Imposing the extremum energy condition, i.e., $dE/du_n = 0$, we obtain $u_n = A_n |\psi_n|^2 / K$. Inserting this back into Eq. (7.3), we get

$$i\frac{d\psi_n}{dt} = \epsilon_n \psi_n - J[\psi_{n-1} + \psi_{n+1}] - (A_n^2/K)|\psi_n|^2 \psi_n + \sum_p |(A_p^2/2K)\psi_p|^4. \quad (7.4)$$

This last step represents a departure from the Holstein adiabatic approach being valid only in the *opposite* limit, i.e., that where the vibrational degrees of freedom adjust rapidly to the excitonic motion. In this anti-adiabatic or *diabatic* limit,

Fig. 7.1 Time evolution of $\psi_n(t)$ for a lattice of $N = 41$ sites with periodic boundary conditions, $V = 1$, uniform nonlinearity $\chi = 1$, and a localized initial condition

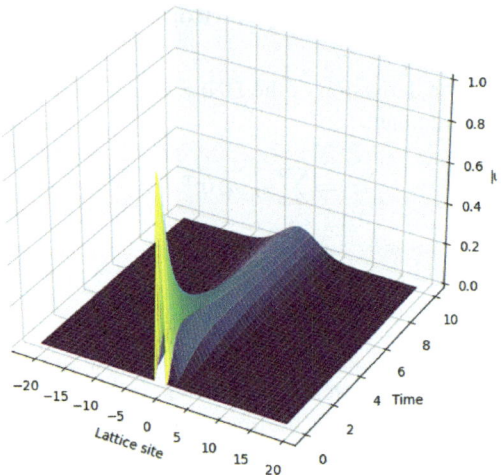

it is still possible to retain the dynamics in the original Eq. (7.3). The quantity $\sum_p (A_p^2/2K)|\psi_p|^4$ represents the total vibrational energy. If we measure energies with respect to this background value, we arrive to an effective nonlinear equation for the amplitude $\psi_n(t)$:

$$i\frac{d\psi_n}{dt} = \epsilon_n \psi_n - J[\psi_{n-1} + \psi_{n+1}] - (A_n^2/K)|\psi_n|^2 \psi_n. \quad (7.5)$$

This closed nonlinear equation describes the effective motion of the "polaron" in the aforementioned anti-adiabatic limit. The "time step" dt in the time derivative should be understood as short compared to the time scale of the "bare exciton motion" (proportional to $1/J$) but long compared to the fast vibrational motion (proportional to $1/K$). We conclude that the regime of validity of DNLS in the context derived above is in the range where $|J| \ll \hbar\Omega_n$, with Ω_n being the frequency of any of the local Einstein oscillators.

Using $V = -J$ and $\chi_n = A_n^2/K$, we arrive at the form of the DNLS equation already stated in Eq. (7.1). Below we include a general Python code in Jupyter notebook format that integrates the DNLS equation with uniform nonlinearity and a localized initial condition. In the output we plot the norm and the actual probability amplitude ψ_n as in Fig. 7.1.

```python
import numpy as np
import matplotlib.pyplot as plt

# Parameters
N = 41   # Number of lattice sites
x = np.arange(-N//2, N//2)   # Lattice positions
T = 10.0   # Total simulation time
dt = 0.01   # Time step
```

7.2 Diabatic Approximation

```python
timesteps = int(T / dt)  # Number of time steps

C = 1.0    # Coupling constant
chi = 1.0  # Nonlinearity strength

# Initial condition: all zero except at site 0
psi_0 = np.zeros(N, dtype=complex)
psi_0[N // 2] = 1.0
# Initial probability amplitude at the center site

# Define the DNLS equation as a function
def dnls_rhs(psi, C, chi):
    psi_shift_left = np.roll(psi, 1)
    psi_shift_right = np.roll(psi, -1)

    # DNLS equation with periodic boundary conditions
    dpsi_dt = -C * (psi_shift_left + psi_shift_right) +
        chi * np.abs(psi)**2 * psi
    return dpsi_dt

# Crank-Nicolson method for time evolution
def crank_nicolson(psi, C, chi, dt):
    # Half-step advance with explicit Euler (for prediction)
    psi_half = psi + 0.5 * dt * dnls_rhs(psi, C, chi)

    # Full-step advance using trapezoidal rule (implicit)
    psi_next = psi + dt * 0.5 * (dnls_rhs(psi, C, chi) +
        dnls_rhs(psi_half, C, chi))

    # Return the updated wavefunction
    return psi_next

# Time evolution
psi_sol = np.zeros((N, timesteps), dtype=complex)
psi_sol[:, 0] = psi_0

# Iterate over timesteps
for i in range(1, timesteps):
    psi_sol[:, i] = crank_nicolson(psi_sol[:, i-1], C, chi, dt)

    # Normalize at each step to ensure norm conservation
    norm = np.sqrt(np.sum(np.abs(psi_sol[:, i])**2))
    psi_sol[:, i] /= norm

# Calculate and print the norm at each time step to verify norm
 #   conservation
norms = np.sum(np.abs(psi_sol)**2, axis=0)
for i, norm in enumerate(norms):
    print(f"Time: {i*dt:.3f}, Total Norm: {norm:.6f}")

# Plotting the norm over time to check its conservation
plt.figure(figsize=(10, 6))
plt.plot(np.arange(timesteps) * dt, norms, label='Total Norm')
plt.xlabel('Time')
```

```
plt.ylabel('Total Norm')
plt.title('Norm Conservation Over Time (Crank-Nicolson)')
plt.grid(True)
plt.show()

# Plotting the results in 3D
fig = plt.figure(figsize=(10, 7))
ax = fig.add_subplot(111, projection='3d')

# Ensure the x array matches the number of lattice sites (N)
T_grid, X_grid = np.meshgrid(np.arange(timesteps) * dt, x)

# Plot the surface
ax.plot_surface(X_grid, T_grid, np.abs(psi_sol)**2, cmap='viridis')

ax.set_xlabel('Lattice site')
ax.set_ylabel('Time')
ax.set_zlabel('||\textsuperscrit 2')
ax.set_title('DNLS Equation: Time Evolution with Crank-Nicolson')

plt.show()
```

7.3 General Properties of DNLS

In the previous section we showed how the DNLS equation can be motivated in a solid-state context. In optics context, on the other hand, the nonlinear term models genuine properties of the medium and the equation describes wave motion in coupled nonlinear waveguides. In this case ψ_n is the amplitude coefficient in an expansion of the electromagnetic field in terms of the wave normal modes in the waveguide. The coupling enables optical power sharing and energy exchange among the waveguides. The nonlinear nature of the materials in each waveguide or coupler can cause a trapping of power in one of the waveguides. The wave propagation takes place along the wave guides, and the self-trapping phenomenon is a function of a space variable instead of a time on. In other words, the time t in Eq. (7.1) describes propagation in space along each waveguide [6, 7].

In order to address the optics problem we rewrite the DNLS equation in a slightly different form. Let us define $\psi_n \equiv \tilde{c}_n$, while changing the normalization condition to $\sum_n |\psi_n|^2 = P$, where P is the total electromagnetic power injected into a waveguide system; we recall that $P = 1$ in the exciton problem. The DNLS equation can then be written as

$$i\frac{d\tilde{c}_n}{dt} = \epsilon_n \tilde{c}_n + V(\tilde{c}_{n-1} + \tilde{c}_{n+1}) - \chi_n |\tilde{c}_n|^2 \tilde{c}_n \tag{7.6}$$

and

7.3 General Properties of DNLS

$$\sum_p |\tilde{c}_p|^2 = P. \tag{7.7}$$

In the homogenous problem where $\chi_n \equiv \chi$, we have similarly $\tilde{\chi}_n \equiv \tilde{\chi}$. While in the optics problem we are interested in the critical power P_{cr} that induces self-focusing *in space*, in the condensed matter problem we focus on the critical nonlinearity χ_{cr} that causes self-trapping in time. The two problems are identical however since by simply rescaling the variables and defining $c_n = \tilde{c}_n/\sqrt{P}$ and $\tilde{\chi}_n P = \chi_n$ Eq. (7.6) transforms to

$$i\frac{dc_n}{dt} = \epsilon_n c_n + V(c_{n-1} + c_{n+1}) - \chi_n |c_n|^2 c_n \tag{7.8}$$

$$\sum_p |c_p|^2 = 1. \tag{7.9}$$

Equation (7.8) is identical to Eq. (7.1) only with variables c_n instead of ψ_n. In other words the input power scales the nonlinearity parameter, while the evolution equation remains the same. What is very interesting in optics is that the nonlinear term depends on the input power and thus may be relatively easily tuned. This is in marked contrast to the condensed matter problem where nonlinearity is intrinsic since it depends entirely on the electron-phonon coupling.

In the DNLS equation we have two conserved quantities; one is the norm $\sum_n |\psi_n|^2$,[2] while the second is the energy H given by

$$H = \sum_n \epsilon_n |\psi_n|^2 + V \sum_n (\psi_{n-1} + \psi_{n+1})\psi_n^* - \frac{1}{2}\sum_n \chi_n |\psi_n|^4. \tag{7.10}$$

If we assume that $\epsilon_n = \epsilon$ and $\chi_n = \chi$ for all n and perform the transformation $\psi_n \to \exp(-i\epsilon t)\psi_n$, we obtain

$$i\frac{d\psi_n}{dt} = V(\psi_{n-1} + \psi_{n+1}) - \chi |\psi_n|^2 \psi_n. \tag{7.11}$$

This is the simpler version of the DNLS equation when all units, molecular or optical, are identical in terms of both the linear diagonal term and the nonlinear parameter.

Equation (7.11) represents actually two equations for $\text{Re}(\psi_n)$ and $\text{Im}(\psi_n)$ or equivalently for ψ_n and ψ_n^*; the equation for the latter is

$$i\frac{d\psi_n^*}{dt} = -V(\psi_{n-1}^* + \psi_{n+1}^*) + \chi |\psi_n|^2 \psi_n^*. \tag{7.12}$$

---

[2] We will use the original ψ_n variable from here on with unit normalization.

In principle V and χ can have either sign depending on the physical problem at hand. For the mathematical study it suffices to consider $\chi > 0$ and V with either sign [5].

7.4 The Degenerate Nonlinear Dimer

The simplest DNLS equation unit that can be analyzed is the degenerate dimer. This constitutes of two sites, 1 and 2, with the same energies that for convenience can be taken to be equal to 0, i.e., $\epsilon_1 = \epsilon_2 = 0$, and identical nonlinearity parameters, i.e., $\chi_1 = \chi_2 = \chi$. The resulting equations are

$$i\frac{d}{dt}\psi_1 = V\psi_2 - \chi|\psi_1|^2\psi_1 \tag{7.13}$$

$$i\frac{d}{dt}\psi_2 = V\psi_1 - \chi|\psi_2|^2\psi_2. \tag{7.14}$$

The degenerate dimer described through the set of Eqs. (7.13) and (7.14) is a unique system since it is the simplest DNLS unit that is fully integrable. As a result, many of the basic properties induced by nonlinearity in the exciton transfer can be understood from this system.

7.4.1 Density Matrix Equations

The nonlinear dimer equations are complex and nonlinear, and it is appropriate to turn them into a corresponding set of equations where the unknown functions are real. Since the total probability is a conserved quantity, i.e., $|\psi_1|^2 + |\psi_2|^2 = 1$, we expect the system of Eqs. (7.13) and (7.14) to reduce to three real ode's. A physically motivated way to rewrite the equations for real variables is to proceed through the density matrix. This is defined as $\rho_{mn} = \psi_m \psi_n^*$, where the star * denotes complex conjugation. For the nonlinear dimer the density matrix is a 2×2 matrix where the diagonal elements denote occupation probabilities and the off-diagonal ones contain phase information. In order to turn Eqs. (7.13) and (7.14) into density matrix equations, we form the time derivative of the (m, n) element of the density matrix for $m, n = 1, 2$ as $\dot{\rho}_{mn} = \dot{\psi}_m \psi_n^* + \psi_m \dot{\psi}_n^*$ and substitute the derivative terms from the equations to the right-hand side. We obtain

$$\dot{\rho}_{11} = iV(\rho_{12} - \rho_{21}) \tag{7.15}$$

$$\dot{\rho}_{22} = -iV(\rho_{12} - \rho_{21}) \tag{7.16}$$

$$\dot{\rho}_{12} = +iV(\rho_{11} - \rho_{22}) + i\chi(\rho_{22} - \rho_{11})\rho_{12} \tag{7.17}$$

7.4 The Degenerate Nonlinear Dimer

$$\dot{\rho}_{11} = -iV(\rho_{11} - \rho_{22}) + i\chi(\rho_{11} - \rho_{22})\rho_{21}. \tag{7.18}$$

This set of equations is clearly nonlinear, but some hints about the dynamics are already discernible. To expose this we may now introduce further a new set of variables, viz. $s = \rho_{11} + \rho_{22}$, $p = \rho_{11} - \rho_{22}$, $q = i(\rho_{12} - \rho_{21})$, and $r = \rho_{12} - \rho_{21}$ that are linear combinations of the diagonal and off-diagonal matrix elements, respectively. Simple algebra leads to

$$\dot{p} = 2Vq \tag{7.19}$$

$$\dot{q} = -2Vp - \chi pr \tag{7.20}$$

$$\dot{r} = \chi pq, \tag{7.21}$$

while the equation for s is the normalization, viz. $s = 1$. This set of differential equations is much simpler; we may use Eq. (7.19) in Eq. (7.21) and obtain an equation for variables r and p as follows:

$$\frac{d}{dt}r = \frac{\chi}{2V}\frac{d}{dt}(p\dot{p}). \tag{7.22}$$

Equation (7.22) can be fully integrated since its RHS is proportional to the time derivative of p^2; it has the solution

$$r(t) = (r_0 - \frac{\chi}{4V}p_0^2) + \frac{\chi}{4V}p^2, \tag{7.23}$$

where r_0 and p_0 are the initial values of these variables. To continue, we differentiate once Eq. (7.19) and substitute $\dot{q}$ from Eq. (7.23) while replacing also the variable r with the value in Eq. (7.23). After these steps we arrive at the second order equation for the probability difference p:

$$\ddot{p} = -\left[(2V)^2 + 2V\chi r_0 - \frac{\chi^2}{2}p_0^2\right]p - \frac{\chi^2}{2}p^3. \tag{7.24}$$

We observe that Eq. (7.24) is an equation similar to Newton's second law of motion where the probability difference $p \equiv p(t)$ plays the role of the position of a particle; on the LHS we have the acceleration term while on the RHS the force that enables the motion. By multiplying both sides of Eq. (7.24) with $\dot{p}$ and integrating, we obtain an "energy conservation equation" in the form

$$\dot{p}^2 + V(p) = E_0 \tag{7.25}$$

$$V(p) = \left[(2V)^2 + 2V\chi r_0 - \frac{\chi^2}{2}p_0^2\right]p^2 + \frac{\chi^2}{4}p^4 \tag{7.26}$$

$$E_0 = \dot{p}_0^2 + \left[(2V)^2 + 2V\chi r_0 - \frac{\chi^2}{2}p_0^2\right]p_0^2 + \frac{\chi^2}{4}p_0^4. \tag{7.27}$$

In order to gain further intuition in the nonlinear dimer dynamics, it is judicious to study different specific initial conditions instead going directly to the general solution.

7.4.2 Localized Initial Conditions

We consider the case where the excitation is initially localized completely on the first site; this corresponds to the initial conditions $p_0 = 1$, $q_0 = r_0 = 0$. We may understand qualitatively the various features of the motion if we look at the effective potential $V(p)$ and the corresponding classical, effective equation of motion [8].

7.4.2.1 Effective Classical Particle Motion

For the initially localized conditions the particle is placed completely initially on the first site, while no complex phases are involved. The set of dynamic Eqs. (7.25, 7.26, 7.27) simplifies to

$$\dot{p}^2 + V(p) = E_0 \tag{7.28}$$

$$V(p) = \left[(2V)^2 - \frac{\chi^2}{2}\right]p^2 + \frac{\chi^2}{4}p^4 \tag{7.29}$$

$$E_0 = (2V)^2 - \frac{\chi^2}{4}, \tag{7.30}$$

where $\dot{p}_0 = 2Vq_0 = 0$. To simplify further the equations, we rescale time to $\tau = 2Vt$ and introduce the variable $\zeta = \chi/4V$; we obtain

$$\left(\frac{dp}{d\tau}\right)^2 + V(p) = \epsilon_0 \tag{7.31}$$

$$V(p) = \left(1 - 2\zeta^2\right)p^2 + \zeta^2 p^4 \tag{7.32}$$

$$\epsilon_0 \equiv \frac{E_0}{(2V)^2} = 1 - \zeta^2. \tag{7.33}$$

Equations (7.31), (7.32), and (7.33) describe the evolution of a particle with mass equal to $1/2$ that executes dynamics in a quartic potential. To make the representation simpler, it is preferable to absorb the nonlinear term $1 - \zeta^2$ in the

7.4 The Degenerate Nonlinear Dimer

effective potential so that the effective initial energy is independent of ζ and equal to 0. We define the new potential $U(p)$ as follows:

$$U(p) = V(p) + \zeta^2 - 1 \equiv \zeta^2 p^4 + \left(1 - 2\zeta^2\right) p^2 + \zeta^2 - 1. \tag{7.34}$$

With definition of Eq. (7.34) the conservation of energy for the effective particle becomes

$$\left(\frac{dp}{d\tau}\right)^2 + U(p) = 0. \tag{7.35}$$

We plot the effective potential $U(p)$ as a function of p for various values of the nonlinearity parameter ζ in Fig. 7.2. The red bullet on the right denotes the initial condition while the horizontal dashed line the initial particle energy compatible with a specific nonlinearity value. In this figure we have different potential curves, each corresponds to the effective classical potential for the corresponding ζ-value.

Let us follow the dimer dynamics through this potential picture. In the linear dimer for $\zeta = 0$ the potential is purely quadratic, and a particle starting at $p_0 = 1$ simply performs oscillatory motion between the two extremes at $p = \pm 1$. The dynamics is purely trigonometric and the quantum particle simply oscillates completely between the two energy states available to if with period equal to $2VT = 2\pi$, i.e., $T = \pi/V$. Once ζ becomes nonzero, the potential $U(p)$ starts deforming and becomes more shallow without however losing the contact points at $p = \pm 1$ with the horizontal constant energy line. For small ζ's the dynamics is qualitatively similar to the linear motion, i.e., periodic dynamics between the two extremes. Since the potential is more shallow, the kinetic energy at the bottom is smaller and, as a result, the dynamics becomes slower.

For the nonlinearity value $\zeta = 1/\sqrt{2} \approx 0.707$, the second derivative of $U(p)$ at $p = 0$ becomes zero, and the bottom of the potential becomes flat, as in Fig. 7.2. It is clear that the particle dynamics is very slow for this value of nonlinearity as the particle crosses from positive to negative values of p. As ζ increases past this value, the quadratic term in the potential becomes negative, and a barrier develops at $p = 0$. The effective particle has to overcome this effective barrier—this means that the quantum particle dynamics becomes even slower as it tries to tunnel from the first site to the second one. The period of the periodic motion between the two extremes reduces as the nonlinearity value increases and actually becomes infinite at $\zeta = 1$. This marks the *self-trapping transition* where the quantum particle cannot cross completely to the second site, and it simply reaches equipartition of probability. In the effective particle picture, the maximum of potential $U(p)$ at $p = 0$ reaches the value "zero," i.e., $U(0) = 0$, and due to energy conservation; the particle that starts from $p = 1$ can only reach asymptotically this maximum. The motion ceases to be periodic for this nonlinearity value and becomes hyperbolic. Further increase of ζ shifts the maximum $U(0)$ to values larger than 0, and the motion of the particle is again periodic yet incomplete. The quantum particle executes oscillations, but

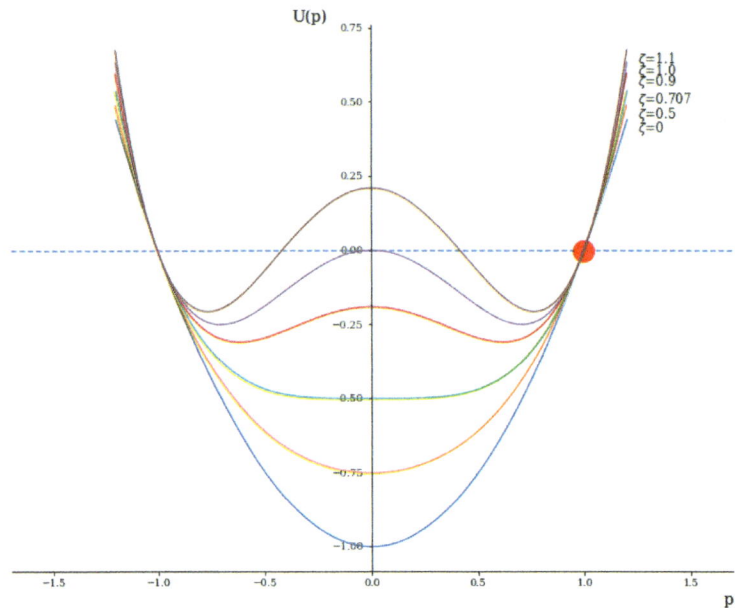

Fig. 7.2 Effective potential motion for self-trapping with localized initial condition. On the abscissa the probability difference p, in the ordinate the effective potential $U(p) = V(p) + \zeta^2 - 1$. The horizontal line demarks the initial energy while the red bullet the initial state

it can reach over to the other side only partially to a degree that depends on the nonlinearity value. This behavior is reminiscent of a linear but not degenerate dimer. As nonlinearity increases without limit, the particle localization on the initial site becomes more pronounced.

The different regimes of the effective particle dynamics are portrayed through the nonlinear potential in Fig. 7.2. While close to the linear regime the quantum particle executes complete oscillations between the two sites, yet the period of oscillation increases as the nonlinearity increases. The more "dressed with phonons" is the particle, the more "sluggish" it becomes, yet the motion is quite similar to the pure periodic motion of the linear degenerate dimer. The self-trapping transition occurs when the period of oscillatory motion becomes infinite; it takes infinite time for the two sites to reach equipartition. Past this point the motion becomes again periodic but incomplete, since the particle cannot reach fully the second site. The motion now is very reminiscent of the linear nondegenerate dimer where the two sites are not at resonance. As the nondegeneracy increases, the sites become more and more out of resonance and thus the transfer to the second site less and less pronounced.

7.4.2.2 Exact Time Dependent Solution

In order to obtain the exact time dependent solution, one has to solve the differential Eq. (7.35) or Eq. (7.24) for the appropriate initial conditions. We can write formally the solution of the former as

$$\tau = \int_1^p \frac{dp'}{\sqrt{-U(p')}} = \int_1^p \frac{dp'}{\sqrt{-[\zeta^2 p^4 + (1 - 2\zeta^2) p^2 + \zeta^2 - 1]}}. \quad (7.36)$$

We notice that by construction the values $p = \pm 1$ are roots of the potential $U(p)$; as a result we can express it in the simpler form

$$U(p) = (p^2 - 1)(\zeta^2 p^2 + 1 - \zeta^2). \quad (7.37)$$

Making the substitution $p = \cos\theta$, we have

$$\tau = \int_0^{\cos^{-1}\theta} \frac{-\sin\theta' d\theta'}{\sqrt{\sin^2\theta'(1 - \zeta^2 \sin^2\theta')}} = -\int_0^{\cos^{-1}\theta} \frac{d\theta'}{\sqrt{1 - \zeta^2 \sin^2\theta'}}. \quad (7.38)$$

Equation (7.39) is simply the definition of an elliptic integral of the first kind—its inverse is a Jacobian elliptic function, specifically the Jacobian elliptic cosine. Finally,

$$p(t) = cn[\tau|\zeta] \equiv cn[2Vt|\frac{\chi}{4V}]. \quad (7.39)$$

The exact solution of the nonlinear dimer for the localized initial condition is then expressed through the Jacobi cosine function;[3] the latter contains all the precise dynamics explained qualitatively previously [8, 9]. In the function $cn[\tau|\zeta]$, τ is the independent variable, while ζ is the *elliptic modulus*. In the range $0 \leq \zeta < 1$ the elliptic cosine is periodic with extreme values ± 1. For $\zeta = 0$ the function reduces to the usual trigonometric cosine, viz. $cn[\tau|0] \equiv \cos(\tau)$, while for $\zeta = 1$ it reduces to a hyperbolic function, viz. $cn[\tau|1] \equiv sech(\tau)$. This is the precise functional form of the self-trapping transition. For values $\zeta > 1$, we may apply a transformation and express the Jacobi cosine as $cn[\tau|\zeta] \equiv dn[\zeta\tau|1/\zeta]$, where dn is another Jacobian function.[4] It is also periodic, but the amplitude of its oscillation depends on ζ, similar to the qualitative analysis. We have the identity $dn^2 + \zeta^2 sn^2 = 1$, where sn is the Jacobian sine; $sn^2 + cn^2 = 1$ similar to the corresponding trigonometric functions. This mathematical behavior in time is seen in Fig. 7.3 where the self-trapping transition is clearly visible.

---

[3] An alternative notation of the elliptic cosine is $cn(u, k)$ and similarly for the rest of the Jacobian functions.
[4] This Jacobi transformation is $cn[x|k] \equiv dn[kx|1/k]$.

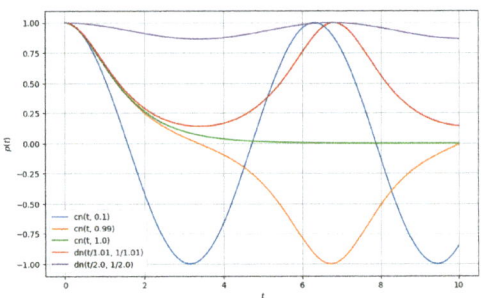

Fig. 7.3 Elliptic function solution of the initially localized nonlinear dimer. The self-trapping transition occurs for elliptic modulus equal to 1, while for larger values of the nonlinearity we see the incomplete oscillations between the two sites that are characteristic of self-trapping

7.4.3 Weierstrass Solution

After we have understood the dimer self-trapping transition qualitatively or through the Jacobi functions, it is instructive to express the degenerate dimer solution in terms of the Weierstrass elliptic function. In contrast to the Jacobi elliptic function approach, the Weierstrass solution is not intuitive; however the method is more straightforward and produces additional insight. We write the energy Eqs. (7.34) and (7.35) as

$$\left(\frac{dp}{d\tau}\right)^2 = -\zeta^2 p^4 + \left(2\zeta^2 - 1\right) p^2 + 1 - \zeta^2 \equiv f(p) \tag{7.40}$$

and introduce the standard polynomial function $f(p)$ in the form

$$f(p) = a_0 p^4 + 4a_1 p^3 + 6a_2 p^2 + 4a_3 p + a_4, \tag{7.41}$$

where $a_0 = -\zeta^2$, $a_1 = 0$, $a_2 = 2\zeta^2 - 1$, $a_3 = 0$, $a_4 = 1 - \zeta^2$. The formal solution of Eq. (7.40) or, equivalently, Eq. (7.35) is

$$p(\tau) = p_{sr} + \frac{f'(p_{sr})}{4} \frac{1}{\wp(\tau; g_2, g_3) - \frac{1}{24} f''(p_{sr})}, \tag{7.42}$$

where p_{sr} is a simple root of $f(p)$, primes denote differentiation with respect to p, while $\wp(\tau; g_2, g_3)$ is the Weierstrass elliptic function with invariants g_2, g_3. Using the simple root $p_{sr} = 1$, we have the explicit solution for the fully localized initial condition:

$$p(\tau) = 1 - \frac{1}{2} \frac{1}{\wp(\tau; g_2, g_3) + \frac{1+4\zeta^2}{12}}. \tag{7.43}$$

The values of the invariants are given by the following general expressions:

$$g_2 = a_0 a_4 - 4a_1 a_3 + 3a_2^2 \tag{7.44}$$

7.4 The Degenerate Nonlinear Dimer

$$g_3 = a_0 a_2 a_4 + 2a_1 a_2 a_3 - a_2^3 - a_0 a_3^2 - a_1^2 a_4, \tag{7.45}$$

which specialize in the present case as follows:

$$g_2 = -\zeta^2(1-\zeta^2) + 3(\zeta^2-1)^2 = \tag{7.46}$$

$$\frac{4}{3}\zeta^4 - \frac{4}{3}\zeta^2 + \frac{1}{12} \tag{7.47}$$

$$g_3 = -\zeta^2(2\zeta^2-1)(1-\zeta^2) - (2\zeta^2-1)^3 = \tag{7.48}$$

$$\frac{8}{27}\zeta^6 - \frac{4}{9}\zeta^4 + \frac{5}{36}\zeta^2 + \frac{1}{216}. \tag{7.49}$$

The modular discriminant $\Delta = g_2^3 - 27g_3^2$ has the following simple form:

$$\Delta = \frac{1}{16}z^4 - \frac{1}{16}z^2 = \frac{1}{16}\zeta^2(\zeta^2 - 1). \tag{7.50}$$

The conditions $\Delta \neq 0$ or $\Delta = 0$ with $g_2 > 0$ and $g_3 > 0$ lead to periodic solutions for the dimer while $\Delta = 0$ with $g_2 \geq 0$ and $g_3 < 0$ to nonperiodic ones.

The discriminant Δ determines the limiting behaviors of the full time dependence of the degenerate dimer. We have a number of special cases: When $\zeta = 0$, we obtain $\Delta = 0$, while $g_2 = 1/12 > 0$, $g_3 = 1/216 > 0$. This is the case of the linear degenerate dimer with the known trigonometric time evolution. The limiting form of the Weierstrass function in this case is

$$\wp(\tau, 1/12, 1/216) = -\frac{1}{12} + \frac{1}{4}\csc^2(\frac{\tau}{2}), \tag{7.51}$$

where the cosecant trigonometric function csc is the reciprocal of the sin function.

The second case where $\Delta = 0$ is for the value $\zeta = 1$ (assuming positive nonlinearity); for this value $g_2 = 1/12 > 0$ and $g_3 = -1/216 < 0$ leading to

$$\wp(\tau, 1/12, -1/216) = \frac{1}{12} + \frac{1}{4}\csch^2(\frac{\tau}{2}), \tag{7.52}$$

where similarly the hyperbolic cosecant $csch$ is the inverse of the hyperbolic sine $sinh$.

Substituting these two limiting forms back to the general solution given in the Eq. (7.43), we have, respectively,

$$p(\tau) = 1 - \frac{1}{2} \frac{1}{\wp(\tau; 1/12, 1/216) + \frac{1+4\zeta^2}{12}} = 1 - \frac{2}{\csc^2(\tau/2)} = 1 - 2\sin^2(\frac{\tau}{2}) = \cos(\tau) \tag{7.53}$$

Table 7.1 Analysis of the Weierstrassian invariants for different nonlinearity ranges

ζ	g_3	Δ	ω_1	ω_3
$(0, \frac{\sqrt{2}}{2})$	$+$	$-$	Ω	$-\frac{\Omega}{2} + \Omega'$
$(\frac{\sqrt{2}}{2}, 1)$	$-$	$-$	$\|\Omega\| + i\frac{\Omega}{2}$	$-i\Omega$
$(1, \frac{1}{2}\sqrt{\frac{3}{2}\sqrt{2}+2})$	$-$	$+$	$\|\omega'\|$	$-i\omega$
$(\frac{1}{2}\sqrt{\frac{3}{2}\sqrt{2}+2}, \infty)$	$+$	$+$	ω	ω'

Table 7.2 Table of special nonlinearity values and the corresponding Weierstrassian parameters

ζ	g_3	Δ	ω_1	ω_3
0	$\frac{1}{216}$	0	π	$i\infty$
$\frac{\sqrt{2}}{2}$	0	$-\frac{1}{64}$	$\frac{(4-4i)\Gamma(\frac{5}{4})^2}{\sqrt{\pi}}$	$\frac{(4+4i)\Gamma(\frac{5}{4})^2}{\sqrt{\pi}}$
1	$-\frac{1}{216}$	0	$-i\pi$	∞
$\frac{1}{2}\sqrt{\frac{3}{2}\sqrt{2}+2}$	0	$\frac{1}{512}$	$\frac{4\cdot 2^{3/4}\Gamma(\frac{5}{4})^2}{\sqrt{\pi}}$	$\frac{4i\cdot 2^{3/4}\Gamma(\frac{5}{4})^2}{\sqrt{\pi}}$

and

$$p(\tau) = 1 - \frac{1}{2}\frac{1}{\wp(\tau; 1/12, -1/216) + \frac{1+5}{12}} = 1 - \frac{1}{1 + \frac{\mathrm{csch}^2(\tau/2)}{2}} = \mathrm{sech}(\tau). \tag{7.54}$$

We observe that the well-known limiting cases of pure trigonometric evolution for $\zeta = 0$ and hyperbolic decay for $\zeta = 1$ are easily recovered.

The specifics of the Weierstrass-based time dependent solution can be analyzed through the values of the pair (g_3, Δ); the latter controls also the half frequencies ω_1 and ω_3 of the fundamental parallelogram. Changes in the signs of g_3 and Δ reflect changes in the Weierstrassian dynamics and thus in the nonlinear dimer one. From Table 7.1 we notice four distinct ranges for the nonlinearity parameter that lead to dynamics compatible with the one discussed earlier through the potential method. In Table 7.2 we show the specific values where the motion changes character.

The Weierstrassian analysis of the nonlinear dimer with localized initial condition reveals that for *four* distinct values of the nonlinearity parameter ζ we have changes in the characteristics of the motion. The values $\zeta = 0, 1$ coincide with the ones obtained from the Jacobian approach. The former corresponds to the linear dimer case, while the latter signifies the location of the self-trapping transition. As noted earlier, for the value $\zeta = \sqrt{2}/2 \approx 0.707107$ the effective dimer potential becomes flat at $p = 0$, while for yet larger values a local maximum develops. The dimer motion becomes increasingly more sluggish, a tendency that will culminate at $\zeta = 1$ with the self-trapping transition. The characteristic value $\zeta = \frac{1}{2}\sqrt{2 + \frac{3}{\sqrt{2}}} \approx 1.015052$ occurs, while the dimer is in the self-trapping regime—it demarks a change in the slope of the initial time evolution. The role of these four special values of the nonlinearity parameter ζ is easily explored in phase space.

7.4.4 Photonics

The discussion on the exact dynamics of the degenerate nonlinear dimer showed the presence of two regimes in the time evolution: the "free" regime where the quantum particle executes complete motion between the two sites and the "trapped" one where the oscillations are incomplete. The transition from one to the other marks the self-trapping transition. In the case of two degrees of freedom the exact time dependent solution shows the onset of self-trapping and the transition occurs for some value of ratio $\zeta \equiv \chi/4V$. For localized initial occupation of one site and while $\zeta < 1$, the probability of occupation $\rho(t) \equiv p_1(t) \equiv |\psi_1|^2$ of the same site oscillates in a periodic fashion with a period that increases as the nonlinearity parameter ζ increases. At the critical value $\zeta = 1$, the oscillation period becomes Infinite, and for values of $\zeta > 1$ the particle self-traps. It is instructive to observe the behavior of the time-average site probability of the first cite, viz. $\langle \rho \rangle \equiv \langle p_1(t) \rangle_{t \to \infty}$ in Fig. 7.4. We observe that the time-averaged probability is equal to $1/2$ when the dimer is in the "free" regime, while this value changes abruptly at the self-trapping transition. In the self-trapping regime the time evolution has an average value different than $1/2$; this is a direct manifestation of self-trapping [7].

We comment here, primarily for the optics or photonics problem, on the direct relation of the time averaged quantity $\langle \rho \rangle$ and the complete solution $p_1(t)$. In the dimer case we have (for $p(0) = 1$) the elliptic function solution presented previously; we may calculate from the dimer spectrum the time-averaged probability

$$\langle \rho \rangle = \begin{cases} 0 & \text{if } \zeta < 1, \\ \frac{2}{\pi K(1/\zeta)} & \text{if } \zeta \geq 1, \end{cases} \quad (7.55)$$

where K is the complete elliptic integral of the first kind. In the case of the two nonlinear optical couplers, the quantity of interest is the coupler transmissivity for a given input power. This is equivalent to observing the amount of nonlinearity necessary for a transition from one dimer state to the other at a given *space location* (referred to here as "time"). The time $t_o = \pi/2$ corresponds to half a coupling length in the two coupled waveguide system. In Fig. 7.4 we present the comparison of the exact time dependent and time-averaged quantities. We note that whereas the former increases slowly through the transition, the time-averaged probability

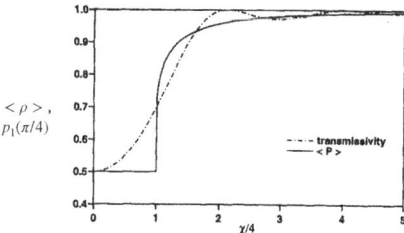

Fig. 7.4 Exact time-averaged probability of occupation of the initially populated site $\langle \rho \rangle$ or $\langle P \rangle$ (continuous curve) and the probability at time $t_0 = \pi/4$, $p_1(\pi/4)$ or "transmissivity" (dashed-dotted curve) for $V = 1$

responds more sensitively to the occurrence of the transition. On the other hand, due to its averaged nature, it does not contain any information pertaining to the gradual oscillation period increases observed in $p(t_0)$.

7.5 The Nondegenerate Nonlinear Dimer

We now turn to the nondegenerate nonlinear dimer that is a system with larger complexity compared to the degenerate one. It is the smallest system where nonlinearity and on-site disorder coexist, and thus its solution gives a perspective for the more general problem of dynamics in nonlinear and disordered lattices. We designate the sites as $n = 1, 2$, the energy difference between the sites with $\Delta = \epsilon_2 - \epsilon_1$, while $p(t)$ is as given previously the probability difference in the occupation of the two sites. We also work with localized initial conditions, viz. $p(0) = |\psi_1(0)|^2 = 1$, that are the simplest yet contain all basic ingredients of the motion. We map the problem to the equivalent classical motion of a particle that executes dynamics in a potential $U(p)$; the equations of motion are

$$\left(\frac{dp(\tau)}{d\tau}\right)^2 + U(p) = 0 \tag{7.56}$$

$$U(p) = (p-1)f(p) \tag{7.57}$$

$$f(p) = a_0 p^3 + a_1 p^2 + a_2 p + a_3 \tag{7.58}$$

$$a_0 = \zeta^2 \tag{7.59}$$

$$a_1 = \zeta(2\delta - \zeta) \tag{7.60}$$

$$a_2 = \delta^2 - \zeta^2 + 1 \tag{7.61}$$

$$a_3 = 1 - (\delta + \zeta)^2 \tag{7.62}$$

with $\tau = 2Vt$, $\zeta = \chi/4V$, and $\delta = (\epsilon_2 - \epsilon_1)/2V$. If p is thought as a position variable, as in the degenerate dimer case, then Eq. (7.56) describes in energy space the motion of a classical particle with mass equal to 2 in a general nonsymmetric quartic potential.

For the initial condition $p(0) = 1$ the total particle energy is equal to 0, and the effective particle starts the motion with zero initial momentum. The coefficients of the quartic potential $U(p)$ depend not only on the nonlinearity ζ but also on the nondegeneracy δ, and thus the dynamics depends strongly on both these parameters. For the localized initial condition one of the roots of the quartic is always $p = 1$. It is therefore sufficient to analyze the roots of the cubic $f(p)$. In Table 7.3 we tabulate three simple limiting expressions of $f(p)$. In the linear degenerate case ($\zeta = \delta = 0$), the cubic degenerates to a linear function with a single root at $p = -1$. In this linear case $U(p)$ has the two roots at $p = 1, -1$, and the solutions are trigonometric.

7.5 The Nondegenerate Nonlinear Dimer

Table 7.3 Limiting cases for the cubic $f(p)$ of Eq. (7.58): Linear degenerate dimer (fist row), linear nondegenerate dimer (second row), and degenerate nonlinear dimer (third row)

δ	ζ	$f(p)$
0	0	$p+1$
δ	0	$(1+\delta^2)p + 1 - \delta^2$
0	ζ	$\zeta^2 p^3 - \zeta^2 p^2 + (1-\zeta^2)p + 1 - \zeta^2$

For the purely linear case with $\zeta = 0$ and $\delta \neq 0$, the negative root of $f(p)$ at $p = -1$ shifts as a function of δ to the value $p = p_\delta > -1$. As the value of the nondegeneracy parameter δ is positive and increases, p_δ moves toward the fixed positive root at $p = 1$, reaching the latter at infinitely large values of δ. This root movement corresponds to increasingly incomplete oscillations between the two sites in the linear nondegenerate dimer.

If we compare this "root movement" with the nonlinear degenerate dimer discussed in the previous section where $\delta = 0$ and $\zeta \neq 0$, we note that in that case the root at $p = -1$ is always *fixed* allowing complete oscillations between the two dimer sites even in the presence of finite nonlinearity ($\zeta < 1$). In this regime of "free" complete oscillations of the degenerate dimer, the cubic $f(p)$ has one real and two complex conjugate imaginary roots whose actual value decreases as the nonlinearity increases. At $\zeta = 1$ the imaginary roots become equal to 0, and the cubic $f(p)$ has *one double root* at $p = 0$. The potential $U(p)$ is now a "fully developed" double well potential, and due to energy conservation the particle launched at $p = 1$ cannot go beyond $p = 0$. The occurrence of the double root in the cubic $f(p)$ denotes the onset of the self-trapping transition. This is seen clearly in Fig. 7.2 of the degenerate dimer potential. For larger values of ζ the potential $U(p)$ has four real roots, two of which are in between the fixed roots at $p = 1, -1$. In this case the classical particle executes incomplete oscillations, and the dimer is in a self-trapped state.

When nondegeneracy and nonlinearity are both present, the "root dynamics" effects mix. Nondegeneracy shifts the negative root at $p = -1$ toward the larger values at $p = p_\delta$, while nonlinearity restricts the available space accessible to the particle by introducing the two new roots in between the ones at the extremes of the motion. For very large nondegeneracy, p_δ grows, the motion becomes very restricted, and the effects of nonlinearity and nondegeneracy cannot be separated. For small values of δ, however, the two tendencies due to nonlinearity and nondegeneracy approximately separate. In order to extend the self-trapping idea to nonzero δ, we resort to the effective motion potential picture where it is natural to associate self-trapping with the point at which a *double root* appears in the cubic $f(p)$ or, equivalently, the potential $U(p)$.

When a double root appears, the dynamical space available to the particle becomes drastically restricted, and the dimer falls in a self-trapped state. We note that this condition corresponds to an *abrupt* change in the behavior of the nonlinear nondegenerate dimer in sharp contrast to the gradual change that takes place when only nondegeneracy is present, as in the corresponding linear case.

The condition for the presence of a double root in the cubic $f(p)$ can be found very easily and leads to constraint for ζ and δ:

$$8\delta\zeta^3 + 12\delta^2\zeta^2 - \zeta^2 + 6\delta^3\zeta - 10\delta\zeta + \delta^4 + 2\delta^2 + 1 = 0. \tag{7.63}$$

The solutions of Eq. (7.63) describe the *transition lines* for the self-trapping transition of the nondegenerate dimer [10]. The transition point at $\zeta = 1$ in the degenerate nonlinear dimer case becomes now an entire *transition line* that depends on (δ, ζ). To gain intuition it is instructive to derive the analytical solutions of Eq. (7.63) in the case of small δ. Using Taylor expansion for the nonlinear term and keeping the lowest order terms, we find the equation $\zeta + \delta = 1$; this is the equation for the critical transition line for small nondegeneracies in the (ζ, δ) space. Solving for the critical nonlinearity, i.e., writing $\zeta \equiv \zeta_{cr}$, we have

$$\zeta_{tr} \approx 1 - \delta. \tag{7.64}$$

Equation (7.64) determines the value of the critical nonlinearity as a function of the nondegeneracy but in the limit of small δ close to the degenerate dimer. We observe that the transition point shifts linearly by δ in this small nondegeneracy regime. When $\delta = 0$ we known the result that the self-trapping transition is at $\zeta_{tr} = 1$. We also observe that the transition point depends on the sign of δ and that the transition can occur at *smaller* nonlinearity values for positive nondegeneracies. For $\delta > 0$ the second site has higher energy than the originally populated site and thus effectively acts as a "barrier" that helps the localization of the particle in the first site. When the second site has lower energy, on the other hand, more nonlinearity is necessary before the particle gets trapped.

The complete solution of Eq. (7.63) leads to the phase diagram of Fig. 7.5. The cubic of Eq. (7.63) has a double root for values of δ and ζ corresponding to the continuous line of Fig. 7.5. There is also an alternative solution that is not shown in the figure and is of secondary nature. The continuous line is the transition line for the nondegenerate nonlinear dimer since it corresponds to the presence of a double root in between the roots at $p = 1$ and $p = p_\delta$. As we move across this line (in (δ, ζ) space), the motion of the particle changes abruptly: In the direction of larger nonlinearity we reach the self-trapped states, while in the opposite direction we switch to the almost complete dimer motion of the degenerate dimer. The additional solution of the cubic equation corresponds to an alternative double root of $f(p)$ that forms always close to p_δ [10]. It is related to the change in slope of the potential $U(p)$ at $p = p_\delta$ from negative to positive. This solution gives the location (in (δ, ζ) space) at which the effects of nonlinearity and nondegeneracy cannot be separated any more. For values of δ and ζ beyond this solution, the two effects

7.5 The Nondegenerate Nonlinear Dimer

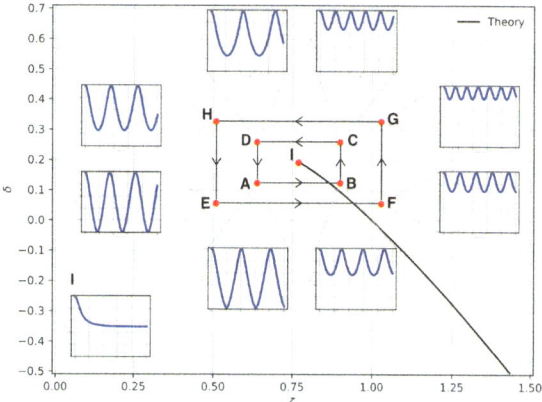

Fig. 7.5 The dynamical domains of the nonlinear nondegenerate dimer. The continuous line corresponds to the transition line for the dimer, while the bullet designates the critical point where the transition line terminates. The paths $A \to B \to C \to D \to A$ and $E \to F \to G \to H \to E$ show two complete paths around the critical point, while in the inserts we plot the precise time dependence in the edge points of the path parallelograms

are completely indistinguishable.[5] The bullet at $(\delta_{cr}, \zeta_{cr}) = (1/\sqrt{27}, 4/\sqrt{27}) \approx (0.19245, 0.76980)$ is a *critical point* at which the two double root lines merge. The presence of this critical point where the transition line terminates allows, as in the cases of ordinary phase transitions, for the possibility of a *continuous path* to the self-trapped regime *without* having to cross the self-trapping transition line, which acts here as a coexistence curve.

The prominent feature in Fig. 7.5 is the thick transition line that terminates in the bullet designated as I. The picture is similar to that of phase transitions with a critical point I. When crossing through the transition line, we change phase, but since the line is terminating we may return to the original state gradually. Let us follow the two paths shown in the figure, starting from the outer one. At A the system is weakly nonlinear with a small nondegeneracy δ; we observe that the time evolution of the probability $p_1(t)$ to be on the initial site oscillates periodically between approximately 1 and 0. The particle thus executes essentially complete oscillations between the two sites. We follow the line $A \to B$ that changes the nonlinearity while keeping the nondegeneracy fixed. At B we are in the self-trapped regime as is shown clearly in the time evolution of $P_1(t)$. Subsequently we increase the nondegeneracy δ while keeping ζ fixed and arrive at the point C. Here the nondegeneracy dominates and with the help of nonlinearity the excitation is fully trapped in the initial site. Reducing the nonlinearity we arrive through the branch

---

[5] The second solution of Eq. (7.63) that also terminates in the same point as the transition line does not appear in the numerics clearly and calls for a proper indicator that will show the difference between linear and nonlinear localization.

$C \to D$ at the point D where the nonlinearity is reduced. We observe that the oscillations are incomplete, but there is substantial transition to the second site. The role of nonlinearity is not dominant; it is the linear nondegeneracy that controls the dynamics. Finally through the branch $D \to A$, we return to the original state without having to cross the transition line. The outer cycle $E \to F \to G \to H \to E$ shows a similar behavior.

The cycles show how disorder and nonlinearity are in cooperation and competition in the nonlinear nondegenerate dimer. The nondegeneracy introduces nonresonance between the sites and thus acts as disorder, while nonlinearity acts as a focusing mechanism. We see in this simple model that the two are not easily separable, since for the particular choice of signs we made they both act in a similar way. The transition line, however, stems from the presence of nonlinearity, while the dynamics on the line is aperiodic. The analytical solutions on the transition line can be obtained by evaluating the integral in Eq. (7.56). It is, however, much easier to obtain results for small nondegeneracy δ. For small δ, the equation for the transition line is $\zeta_{tr} \approx 1 - \delta$ (Eq. 7.64), leading to the following roots (to the first order in δ) for the cubic $f(p)$:

$$p_\delta \approx -1 \tag{7.65}$$

$$p_d \approx \delta, \tag{7.66}$$

where p_d is the double root. The resulting time dependent solution is the following:

$$p(\tau) = \frac{\delta + sech(1-\delta)\tau}{1 + \delta sech(1-\delta)\tau}. \tag{7.67}$$

We note that for $\delta = 0$, Eq. (7.67) reduces to $p(\tau) = sech(\tau)$ in complete agreement with the results of the degenerate nonlinear dimer, while at long times it approaches the double root at $p_d \approx \delta$.

Finally, we may find the explicit time dependence at the critical point of the transition line. At this point where $(\delta_{cr}, \zeta_{cr}) = (1/\sqrt{27}, 4/\sqrt{27}) \approx (0.19245, 0.76980)$, we find a simple algebraic decay [10]:

$$p(\tau) = \frac{1 - \frac{1}{2}(\frac{2}{3})^4 \tau^2}{1 + (\frac{2}{3})^4 \tau^2} \tag{7.68}$$

designating a slow non-oscillatory transition to a state that adjusts to the two parameters. In the critical point the dynamics marks a slow transition to a state where more than half of the probability has "leaked" to the second site. The specific time evolution of the function $p_1(\tau)$ is shown in the lower left part of Fig. 7.5.[6,7]

---

[6] A complete and detailed exposition of the properties of the nonlinear dimer can be found in the recent book of V. M. Kenkre [11]

[7] Very recent analytical work on the nonlinear nondegenerate dimer can be found in [12].

7.6 Conclusion

The DNLS equation is fundamental both in condensed matter physics and in optics and photonics. It forms a discrete set of equations that are coupled and describes wave dynamics in time or in space. Its main feature is the presence of a cubic nonlinearity term that has profound effects in the evolution and leads to self-trapping. This is a dynamic mechanism that can also be controlled in some cases and lead to interesting physical behavior. The general problem with N sites or degrees of freedom cannot be solved analytically; however simple systems can, and the prototypical one is that of the nonlinear dimer. In the purely degenerate case, a self-trapping transition takes place for a certain value of the nonlinearity parameter. This is a strongly initial condition dependent phenomenon with the case of the localized initial condition giving the most characteristic behavior. In this case, for $\zeta_{cr} = 1$ the occupation probability at the initial site changes abruptly, the dynamics becomes that of a hyperbolic secant, and the excitation cannot reach to the second site completely. For even larger values of ζ, the particle becomes trapped in the original site.

The beauty of the nonlinear dimer is that its solution can be expressed in terms of elliptic functions, i.e., functions that generalize the trigonometric and the hyperbolic functions. The analysis thus can be done completely analytically, while it can be supported by very precise numerics. One may use in the analysis either Jacobian elliptic functions that are more intuitive of the Weierstrass elliptic function that can be handled easier mathematically. In either case the analytical results describe the phenomenon of self-trapping very precisely in the degenerate dimer both of a localized initial condition and of the more general conditions. In the latter case we find interesting behavior and for some cases competition with nonlinearity [13]. Finally, the more general case of the nondegenerate nonlinear dimer provides an interesting and very simple framework to discuss the coexistence of nonlinearity with disorder. This system is also solved analytically and in the space of (ζ, δ) presents a phase diagram with interesting features. We now have a transition line separating self-trapped states from fully oscillatory states. The transition line is similar to a coexistence curve in phase transitions and terminates in a critical point. On the line the dynamics is nonperiodic and marks the onset of a complete change in the physical behavior of the dimer system. In the phase diagram we have regions that are more nonlinearity dominated, those that are nondegeneracy dominated and of course the regions where the behavior is mixed. Since the problem of finding whether the source of localization is linearly or nonlinearly dominated is important, the nondegenerate dimer and other similar systems provide significant information in this direction.

7.7 Summary

- The DNLS equation is fundamental in condensed matter physics and in photonics.
- Nonlinearity induces self-trapping that may be assessed analytically in the case of the degenerate nonlinear dimer system.
- In the nondegenerate nonlinear dimer the coexistence of nonlinearity and disorder leads to a terminating transition line that separates different dynamical phases of the dimer.

References

1. J.C. Eilbeck, P.S. Lomdahl, A.C. Scott, The discrete self-trapping equation. Phys. D **16**, 318–338 (1985)
2. A.S. Davydov, Quantum theory of motion of a quasi-particle in a molecular chain with thermal vibrations taken into account. Phys. Status Solidi (b) **138**, 559 (1986)
3. A.C. Scott, Davydov's soliton. Phys. Rep. **217**(1), 1–67 (1992)
4. N.W. Ashcroft, N.D. Mermin, *Solid State Physics* (Brooks/Cole, 1976)
5. D. Hennig, G.P. Tsironis, Wave transmission in nonlinear lattices. Phys. Rep. **307**(5–6), 333 (1999)
6. D.N. Christodoulides, R.I. Joseph, Discrete self-focusing in nonlinear arrays of coupled waveguides. Opt. Lett. **13**, 794–796 (1988)
7. M.I. Molina, G.P. Tsironis, Dynamics of seftrapping in the discrete nonlinear Schrödinger equation. Phys. D **65**, 267 (1993)
8. V.M. Kenkre, D.K. Campbell, Self-trapping on a dimer: time dependent solution of a discrete nonlinear Schrödinger equation. Phys. Rev. B **34**, 4959(R) (1986)
9. V.M. Kenkre, G.P. Tsironis, D.K. Campbell, Energy transfer, self-trapping, and solitons on a nonlinear dimer, in *Nonlinearity in Condensed Matter*, ed. by A.R. Bishop et al. (Springer, 1987)
10. G.P. Tsironis, Dynamical domains of a nondegenerate nonlinear dimer. Phys. Lett. A **173**, 381 (1993)
11. V.M. Kenkre, Interplay of quantum mechanics and nonlinearity. Lecture Notes in Physics (Springer, 2022)
12. J.D. Andersen, V.M. Kenkre, Surprising features of the energy-mismatched nonlinear dimer. Chaos **34**, 043139 (2024)
13. G.P. Tsironis, V.M. Kenkre, Initial condition effects in the evolution of a nonlinear dimer. Phys. Lett. A **127**, 209 (1988)

Chapter 8
Learning Analytical Solutions

Machine Learning for Elliptic Function Solutions

Abstract In this chapter we try to merge machine learning with the nonlinear dynamics problems discussed in the previous chapter. The application of simple machine learning ideas to recover exact DNLS equation results is a significant test for the applicability of the methods. We focus on the nonlinear dimer problem that is completely known analytically and apply machine learning with the aim to recover the self-trapping transition. We show that both for the simple localized initial condition and for the more general solution of the dimer the machine learning method works very efficiently. Furthermore it also produces the transition line of the nondegenerate dimer. The fact that we can obtain known analytical results through machine learning opens up the possibility of further engagement of artificial intelligence in complex nonlinear systems.

8.1 Machine Learning for the DNLS Dimer

The DNLS dimer plays a central role in the application of machine learning inspired techniques since it is one of the very few completely solvable nonlinear systems. Its main feature is the dynamical self-trapping transition, i.e., a bifurcation phenomenon that depends on the strength of the nonlinearity parameter and the initial conditions. As we saw in the previous chapter, in the degenerate dimer case an initially fully localized excitation on one of the sites executes complete oscillatory motion between the two sites, i.e., the probability of occupation of each site becomes 1 periodically in time. Mathematically the solution is described through elliptic functions, while the nonlinearity controls the period of the motion. At a critical value of the nonlinearity the motion becomes aperiodic—this behavior marks the self-trapping transition. For larger values of the nonlinearity the motion becomes incomplete between the two sites, and the most occupation probability stays at the initial site. When the excitation is not fully localized initially, the dynamics is a bit more complex, but there is still a selftrapping bifurcation that depends on the nonlinearity value and the specific initial condition. This behavior is fully understood analytically as well and depends critically on the initial wave function phases.

In the DNLS dimer the two local energies can also be different; in this case we have a nondegenerate dimer that the smallest system where nonlinearity and some form of disorder coexist. In the latter case, we know analytically that there is a self-trapping transition described mathematically as a whole line in the space of nonlinearity and energy nondegeneracy while depends on initial conditions as well. In this model the dynamics is similar for large nonlinearities or large disorder (energy difference between the two sites) since both aspects lead to extreme probability localization in the initial site. On the other hand, for weak disorder the behavior is similar to that of the degenerate nonlinear dimer with a nonlinearity-driven self-trapping transition.

The mathematics underlying the nonlinear dimer is that of elliptic functions and integrals. The solutions are not only mathematically but also physically well understood. The aim now is to recover these solutions with machine learning "ideology." This means to be able to use a data approach and capture the mathematical and physical behavior of the model. If the method works well in the analytically known cases, it can then be extended to the larger realm of nonanalytically solvable cases. Since we use both data and the equations that lead to data, we may view this approach as a physically motivated machine learning [1].

The approach we implement is the following: We first write a Python code and integrate numerically the DNLS dimer equation using a fourth order Runge-Kutta method with an integration step of 0.005. This numerical approach renders good precision data that describe the dimer evolution for different parameters and different initial conditions. In the spirit of physics motivated machine learning we may now introduce a data free loss function that takes into account the physical aspects of the problem and our search for capturing the self-trapping transition. For the degenerate dimer case all we need to do is to compare the probability of the initial site to the probability value $1/2$; if this value is surpassed, then the dimer with an initially localized condition is certainly not in the self-trapped regime.

One critical point in this approach is what is the actual value of the occupation probability of the initial site p_1 that should be compared to the important $1/2$ value. Probability data should be used but which one? If we let the numerical dimer integration procedure run until a certain time T^{max}, collect all the $p_1(t)$ values until that time, and find the *smallest* one, then this would suffice. We designate the time for the lowest p_1 as $T^{min(p_1)}$ and the actual probability for this time moment $p_1(T^{min(p_1)}, \zeta) \equiv |\psi_1(T^{min(p_1)}, \zeta)|^2$ where we also included in the independent variables the normalized nonlinearity variable ζ. After these preliminaries we may now show the physics inspired loss function to be used:

$$\mathcal{L}(T^{min(p_1)}, \zeta) = \left| \frac{1}{2} - p_1(T^{min(p_1)}, \zeta) \right|. \tag{8.1}$$

The initial value of p_1 is equal to 1 for the localized initial condition. As it evolves in the non-self-trapped regime, the excitation transfers completely to the second site, and thus it may reach the minimum value equal to 0 at the first site; in this case the loss function becomes $1/2$. Precisely at the self-trapping transition, the smallest

8.1 Machine Learning for the DNLS Dimer

value for p_1 is $1/2$, and then the loss function is equal to 0. In the self-trapping regime the minimal value of p_1 becomes larger than $1/2$ since the particle cannot perform complete oscillations to the second site, and thus the loss function ranges from 0 and asymptotically to $1/2$. As a result, if we exclude the limiting $1/2$ value, we have self-trapping when $0 < \mathcal{L} < 1/2$ while free motion for $\mathcal{L} = 0$.

This rather simple yet physically motivated definition for the loss function introduces a certain training difficulty for the machine learning model. This is due to the fact that in the range of nonlinearity $0 < \zeta < 1$ the minimum occupation probability of the first site p_1 is always 0, thus leading to a loss function that is *flat* and equal to $1/2$. A flat loss function that is independent of the nonlinearity ζ makes it very hard to update the weights in the machine learning model, and thus the training stops. This behavior is seen in the area with the white background in Fig. 8.1, where the trajectories initialized inside the range of $0 < \zeta < 1$ are stuck around the initial value of ζ.

In order to bypass the perfect flatness of the loss function in the non-self-trapped regime, we make two simple adjustments in the machine learning approach. The first is almost obvious, viz. we start the machine learning procedure from the self-trapped regime, i.e., initialize all trajectories starting with large enough ζ values. This clearly assures that the loss function is not flat at least in the starting ζ-regime. However, using a large nonlinearity value makes the actual value of the loss function also large, and, as a result, in the process of its minimization we need to use relatively large learning rate. Although this accelerates the process, it may at the same time make large changes in the parameter ζ and ultimately lead the process to the flat loss function regime. In order to avoid this problem, we may introduce one more constraint in the procedure that ensures we are always away from the flat $\mathcal{L}$ regime. If in the last step of the training process the occupation probability p_1 is less than $1/2$, signaling that we have entered the flat loss function regime, then the algorithm resets the ζ to each previous value while, at the same time, reducing the learning rate by a factor of 10. This additional criterion ensures that we always approach the critical value of ζ from values $\zeta > 1$ and that by using an adaptive learning rate, we will converge to the critical value. The algorithm was implemented in TensorFlow 2.4/Keras using an Adam optimizer with a custom learning rate [2]. We proceed now in solving the degenerate dimer problem with machine learning first with localized and then with general initial conditions.

8.1.1 Localized Initial Conditions

We present first the machine learning solution of the degenerate nonlinear dimer for the localized initial condition; we show this in Fig. 8.1. In this figure we have on the abscissa the normalized nonlinearity parameter while on the ordinate the minimum value that the probability on the initial site reaches over the course of the numerical integration of the equations. The filled circles denote the initial

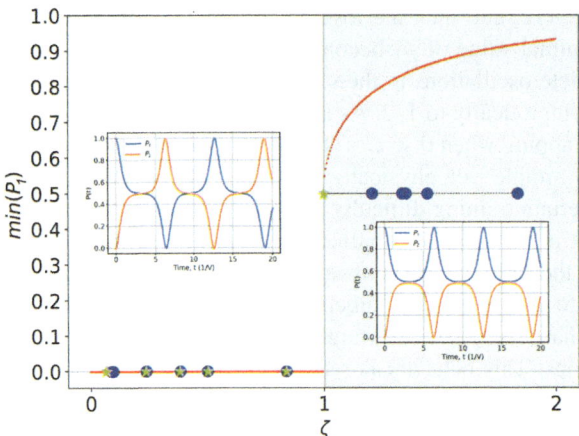

Fig. 8.1 The nonlinear dimer self-trapping transition with ML. The transition is obtained by following the minimal value of the occupation probability at the first site as a function of the scaled nonlinearity ζ. The red dotted line denotes the calculated actual occupation probability $p_1(t)$. The blue bullets denote the initial value of the nonlinearity parameter ζ of ten randomly initialized trajectories (black dashed lines), while the yellow star points to the final value of ζ. This value is either the value that satisfies the desired properties of the system, i.e., self-trapping transition, or a position that the learning process was unsuccessfully stuck. The area with the gray background ($\zeta > 1$) shows the range of values for the nonlinearity that conclude to self-trapping transition. In the inserts we show the time evolution across the transition for $\zeta \to 1_-$ (free motion) and $\zeta \to 1_+$ (self-trapped motion), respectively

value of the nonlinearity used when selecting an ensemble of randomly initialized trajectories (black dashed lines). The plot should be read from right to left: We first fix the value of nonlinearity and then select random trajectories for this value, and through minimization of the loss function we reach at the minimum occupation probability for this specific ζ; the latter is the value shown by the dotted red curve. As the initial ζ reduces, so does the $min(p_1)$ as we readily see by the red curve that reaches the value $1/2$ for $\zeta = 1$ (yellow star point). Further decrease of nonlinearity takes the process to the non-self-trapped regime where the loss function is flat, as explained previously. In the same figure, the gray background area denotes the self-trapping regime, while the subplots present the occupation probability for both sites as a function of time. The plot on the left (white background) is just before the self-trapping transition ($\zeta \to 1_-$) while the one on the right (gray background) just after ($\zeta \to 1_+$).

The results of the machine learning solution of the nonlinear dimer are shown in Fig. 8.1. We notice the precise recovery of the self-trapping transition line that is essentially identical to the analytical one. Furthermore, the procedure captures the exact transition point at $\zeta_{cr} = 1$, i.e., $\chi_{cr} = 4V$ for the perfectly localized initial condition.

8.1.2 General Initial Conditions

When we place the excitation initially on both sites but with different local probabilities, the critical nonlinearity value for self-trapping changes. If we select the more restricted case of real off-diagonal initial density matrix elements, i.e., $q(0) = \rho_{12}(0) - \rho_{21}(0) = 0$, the critical nonlinearity for self-trapping is [3]

$$\zeta_{cr} = \frac{1 \pm (1 - p_0^2)^{1/2}}{p_0^2} \quad (8.2)$$

with $p_0 = \rho_{11}(0) - \rho_{22}(0)$ and where initial real amplitudes may be "in-phase" having the same sign or in "anti-phase" with opposite sign. The former choice leads to the plus sign in Eq. (1.6) while the latter to minus. The critical nonlinearity for self-trapping increases as the amount of initial localization decreases for in-phase motion, while the opposite is true for the out-of-phase motion. This makes sense since the in-phase motion has the tendency of detrapping while the out-of-phase one of retrapping (Fig. 8.2).

In Fig. 8.3, we show the results for general initial conditions with real and positive off-diagonal matrix elements of the density matrix. The continuous line is the analytical result of Eq. (8.2), while the blue bullets present the results of the machine learning search. Both in-phase and anti-phase branches are shown. The agreement between the theory and the machine learning approach is remarkable. This figure shows how well a physics motivated machine learning approach can

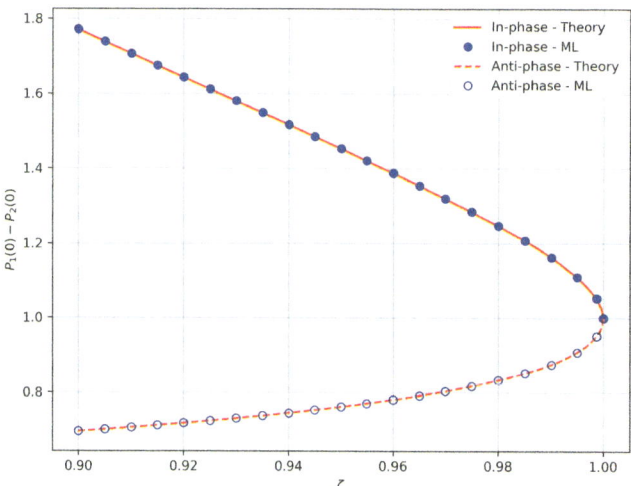

Fig. 8.2 Self-trapping transition with ML for general initial conditions. The solid line is the curve of Eq. (8.2), while the blue bullets were produced with ML. The two branches correspond to in-phase (continuous red line) and anti-phase (dashed red line) initial conditions

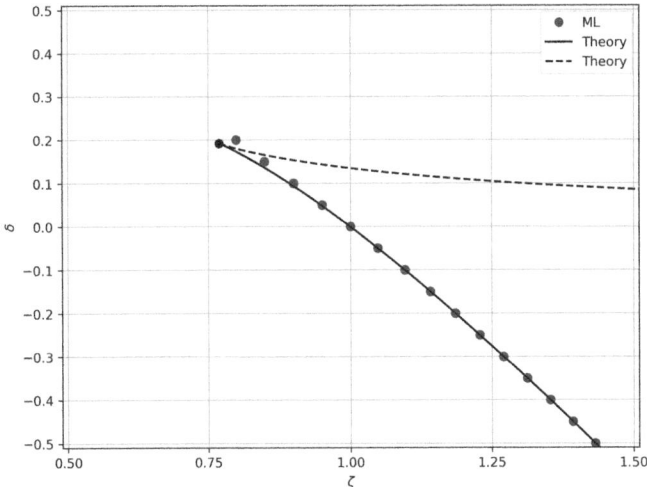

Fig. 8.3 Self-trapping transition with ML for the nondegenerate nonlinear dimer with localized initial condition. The continuous lines are result of analytical calculations while the bullets a result obtained through ML. The solid continuous line represents the self-trapping transition that is a transition to a region of nonlinearly localized states. The dashed line is the second solution of Eq. (7.63) and demarks the onset of strong mixing of linear with nonlinear localization effects. The "tie" region between the continuous and dashed lines corresponds to regime where nonlinear localization is dominant

recover a dynamical phase transition known analytically for a nonlinear dynamical system. Physics can blend very harmoniously with machine learning in complex systems research.

8.2 The Nondegenerate Nonlinear Dimer

As mentioned in the previous chapter the nondegenerate nonlinear dimer is perhaps the simplest "disordered" nonlinear system where the role of disorder plays the energy mismatch between the two sites. Therefore, in addition to nonlinearity and of course the initial conditions we have an additional control parameter, viz. the energy mismatch. In the limit of zero nonlinearity the energy mismatch introduces a linear localization of the excitation in the originally occupied site that is independent of the value of the energy difference between the sites but increases as the latter grows. In the other limit of zero nondegeneracy we have nonlinear localization but this is nonlinearity parameter dependent. When both nonlinearity and nondegeneracy are present, a combined tendency for localization persists having though different features.

The general dimer problem was mapped in the previous chapter into an effective problem of a particle in a potential well. We assumed localized initial conditions

8.2 The Nondegenerate Nonlinear Dimer

and searched for a condition that shows the development of a barrier in the effective potential that prohibits complete motion effective particle motion. With this criterion we found the presence of a transition line in the phase diagram, while in the degenerate dimer we had only a transition point. The analytical results showed the existence of three general regions, viz. a linear-disorder dominated region, a nonlinear dominated region, and a broad mixed phase where both disorder and nonlinearity tendencies coexist. The phase diagram was obtained through analytical means, while the machine learning-based analysis is completely independent of the mathematical approach.

In Fig. 8.3 we compare the analytical with the machine learning results. We observe that the transition line is fully captured by the machine learning approach. The full line denotes the generalized self-trapping transition obtained analytically while the bullets the corresponding numerical results obtained in a way that is similar to the one on the degenerate dimer. The line terminates to a critical point that is also captured by machine learning although with some error. The source of the error is the precise criterion we use for the transition. For relatively small nonlinearities using as criterion the transition of the initial probability to less than $1/2$ is approximately correct. However as we move closer to the critical point, we need to modify this criterion in a way that captures more precisely the dynamics in the effective potential. Since this has not been implemented in this analysis, there is an error between the machine learning result and the analytical one as we approach to the critical region.

In the nondegenerate nonlinear dimer we have disorder and nonlinearity. Broadly speaking, linear localization corresponds to pure Anderson modes while nonlinear localization to discrete breather modes. The dimer provides possibly the simplest system where we may study analytically the competition and coexistence of Anderson modes and discrete breathers. In the phase diagram we may start with an Anderson mode, deform it, and reach a discrete breather mode. This can be done in multiple ways, depending on how we change the parameters δ and ζ. One class of paths crosses the self-trapping line discovered by machine learning, while an alternative class may simply reach the same state without crossing it. The transition from Anderson to discrete breather modes in the present model is similar to a first order phase transition. It is interesting that machine learning can actually capture the coexistence line that separates the two dynamical regimes. The actual time dependent behavior was seen in the previous chapter in Fig. 7.5 where the probability of the initially populated site was shown. Crossing parallel to the ζ-axis is done through the increase in frequency that eventually leads to self-trapping. This is very similar to the degenerate dimer analysis performed previously. The transition to an Anderson mode, on the other hand, is a gradual transition to a more and more localized state induced from the increase of the energy mismatch. In this diagram at the critical separating point (black bullet) the time dependence is algebraic [4].

If we check closely the accuracy of the numerical results obtained with machine learning, we notice that they actually depart from the exact analytical solution when we are close to the terminating point of the transition line. This is hardly a surprise since at this point we have a mixing of localization due to linear and nonlinear parameters. Further work is necessary in order to understand how machine learning can provide the precise critical transitions point—this work will lead also to a better understanding of linear and nonlinear localization mechanisms.

8.3 Conclusion

In this first application of a form of machine learning in complex systems we worked with a specific, integrable configuration of the DNLS equation. This is the nonlinear dimer model that contains only two sites and where the local site energies can be either identical (degenerate dimer) or non-equal (nondegenerate dimer). This simple configuration presents possibly the simplest model where nonlinearity coexists with disorder. We applied machine learning to a number of special cases of the model and found that the numerical solutions essentially coincide with the analytical ones. This is true not only for the localized initial condition but also to more general conditions that involve initial phases. The method depends on the proper choice of the loss function that we chose to be essentially the occupation probability at the non-initially occupied site. This form works well with the specific problem at hand as seen from the results. We find very good agreement of the machine learning results with the analytical ones.

This first successful application of a machine learning approach to an integrable yet complex nonlinear system is extremely encouraging for further use of AI methods in complex systems. If the methods work so well in precisely known cases, then we expect that their domain of applicability extends to other non-integrable systems where we cannot asses their behavior as clearly. The AI approach through learning and subsequent application to unknown cases will be explored further in the following chapters.

8.4 Summary

- The definition of a proper loss function recovers through machine learning the nonlinear dimer self-trapping transition for localized initial condition.
- The method also recovers the general transition for more general initial conditions in the degenerate dimer.
- In the nondegenerate nonlinear dimer the presence of the terminating transition line is also recovered with reasonable accuracy.
- The machine learning method that works very well with the analytical results may be extended in the more general nonanalytical cases of coexistence of nonlinearity and disorder.

References

1. G.E. Karniadakis, I.G. Kevrekidis, L. Lu, P. Perdikaris, S. Wang, L. Yang, Physics informed machine learning. Nat. Rev. Phys. **3**, 422 (2021)
2. G.P. Tsironis, G.D. Barmparis, D.K. Campbell, Dynamical symmetry breaking through AI: The dimer self-trapping transition. Int. J. Mod. Phys. B 2240001 (2021)
3. G.P. Tsironis, V.M. Kenkre, Initial condition effects in the evolution of a nonlinear dimer. Phys. Lett. A **127**, 209 (1988)
4. G.P. Tsironis, Dynamical domains of a nondegenerate nonlinear dimer. Phys. Lett. A **173**, 381 (1993)

Chapter 9
The Targeted Energy Transfer Model

Ultrafast Transfer Through Nonlinear Resonances

Abstract The presence of nonlinearity in the DNLS equation leads to reduction of energy or charge transfer from one site to others. During the chemical processes in photosynthesis, we have an excitation interacting with vibrational modes that tunnels in an ultrafast way. The targeted energy transfer model is based in the DNLS equation and provides a mechanism for fast transfer between sites. The basic feature of the model is that it includes disorder in both the local energy sites and the nonlinearity parameters that are constrained in a specific way. This leads to a nonlinear resonance that enhances transfer and can be analyzed analytically. We use machine learning methodology in order to discover this resonance both in the solvable dimer case as well as the trimer case where there is no analytical solution available. Finding nonlinear resonances through machine learning is quite useful in the study of complex systems.

9.1 The DNLS Donor-Acceptor Dimer

The general form on the DNLS equation introduced in Chap. 7 has two possible sources of disorder where the first stems from the local site energies ϵ_n that may be different for each lattice site n. The second comes from the possible inhomogeneity in the nonlinear terms; the parameter χ_n can be different in each unit labeled by the index n. There are two limits to this equation; the first is when all nonlinearities are zero and the local energies are random; this is the Anderson limit. In this case we have a linear tight binding model with random local energies; it is well-known from the work of P. W. Anderson that at least in one dimension all states are localized. This means that no true motion can take place in this limit. The second case is where all units are identical and thus the local energies are the same, i.e., $\epsilon_n \equiv \epsilon$, but the nonlinearity χ_n is random. Some aspects of this case are known and show that motion is not inhibited by the presence of random nonlinear terms. We will focus here on an intermediate case where both local energies and local nonlinearities are random, but there is a constraint between them. This correlated disorder case leads to enhanced motion compatible also with the dynamics in photosynthetic units in chlorophyll.

The targeted energy transfer model or TET was introduced through a nonlinear nondegenerate dimer system with different nonlinearities in the two sites. We call the first unit "donor," while the second "acceptor" in tune with a chemical picture for the processes involved. The equations that cover the donor-acceptor dynamics are the following:

$$i\frac{d}{dt}\psi_D = -\epsilon\psi_D + V\psi_A + \chi_D|\psi_D|^2\psi_D \tag{9.1}$$

$$i\frac{d}{dt}\psi_A = V\psi_D + \chi_A|\psi_A|^2\psi_A. \tag{9.2}$$

In the TET equations (9.1), (9.2), ψ_D, ψ_A are, respectively, the probability amplitudes for the donor and acceptor molecules. The local energy in the donor is $-\epsilon$, while we take it to be zero in the acceptor site leading to energy mismatch between acceptor and donor equal to $+\epsilon$. The nonlinearity parameters are χ_D and χ_A, respectively, in the two molecules.[1] We note that although we have fixed the energy mismatch there is freedom in the selection of the value of nonlinearities in the two molecules.

9.1.1 Analytical TET Solution

The analytical solution of TET proceeds in the same way as the nonlinear dimer solution; we will assume for simplicity the localized initial solution that places the excitation initially in the donor, i.e., $\psi_D(0) = 1$ and $\psi_A(0) = 0$. Let us skip the initial steps of the calculation and express the four equations for the density matrix:

$$\dot{\rho}_{DD} = iV(\rho_{DA} - \rho_{AD}) \tag{9.3}$$

$$\dot{\rho}_{AA} = -iV(\rho_{DA} - \rho_{AD}) \tag{9.4}$$

$$\dot{\rho}_{DA} = i\epsilon\rho_{DA} + iV(\rho_{DD} - \rho_{AA}) - i(\chi_D\rho_{DD} - \chi_A\rho_A)\rho_{DA} \tag{9.5}$$

$$\dot{\rho}_{AD} = -i\epsilon\rho_{AD} - iV(\rho_{DD} - \rho_{AA}) + i(\chi_D\rho_{DD} - \chi_A\rho_A)\rho_{AD}, \tag{9.6}$$

where $\rho_{DD} = |\psi_D|^2$, $\rho_{AA} = |\psi_A|^2$, $\rho_{DA} = \psi_D\psi_A^*$, and $\rho_{AD} = \psi_A\psi_D^*$. If we define as in the nonlinear dimer case, the auxiliary variables $p = \rho_{DD} - \rho_{AA}$, $q = i(\rho_{DA} - \rho_{AD})$, and $r = \rho_{DA} + \rho_{AD}$, we may express Eqs. (9.3)–(9.6) in the following form:

---

[1] We used a positive sign in front of the nonlinear term that is opposite from the general expression of Eq. (7.1) because here the nonlinearities are parameters to be determined under the TET condition.

9.1 The DNLS Donor-Acceptor Dimer

$$\dot{p} = 2Vq \quad (9.7)$$

$$\dot{q} = -2Vp - \epsilon r + (\chi_D \rho_{DD} - \chi_A \rho_{AA})r \quad (9.8)$$

$$\dot{r} = +\epsilon q - (\chi_D \rho_{DD} - \chi_A \rho_{AA})q. \quad (9.9)$$

In Eqs. (9.7), (9.8), and (9.9) we used the new variables p, q, r, but we retained judiciously the density matrix form in the nonlinear expression. We note the symmetric form of the nonlinear expressions in Eqs. (9.8) and (9.9). Discovering the TET condition using a direct approach and not the dynamical systems method of the original publications [1, 2] proceeds in two steps. In the first we see that if the term $(\chi_D \rho_{DD} - \chi_A \rho_{AA})$ is a constant, then Eqs. (9.8), (9.9) become completely linear. This is easily accomplished by selecting the nonlinearities to have opposite signs, viz. $\chi_D = -\chi_A = \chi$. In this case $\chi_D \rho_{DD} - \chi_A \rho_{AA} = \chi(\rho_{DD} + \rho_{AA}) = \chi$ due to the probability conservation. In this special case, Eqs. (9.7)–(9.9) are written as

$$\dot{p} = 2Vq \quad (9.10)$$

$$\dot{q} = -2Vp - (\epsilon - \chi)r \quad (9.11)$$

$$\dot{r} = (\epsilon - \chi)q. \quad (9.12)$$

We notice that the complex DNLS set of equations has reduced into a simple *linear* set for this choice! This system is equivalent to a linear but not degenerate dimer, and based on the analysis of Chap. 7, it is still not fully resonant. The second step then in the process is done by selecting $\epsilon = \chi$; in this case, the system of Eqs. (9.10)–(9.12) becomes very simple. The third equation becomes $\dot{r} = 0$ and decouples from the other two, while substitution of Eq. (9.11) into Eq. (9.10) after taking its derivative on both sides leads to $\ddot{p} + (2V)p = 0$. This is nothing but the second order equation for the perfectly resonant linear dimer that executes oscillations between the two sites with the fastest rate. For perfectly localized initial conditions, the solution is simply $p(t) = \cos(2Vt)$.

The TET condition

$$\chi_D = -\chi_A = \epsilon \equiv \chi \quad (9.13)$$

determines a unique point in the parameter space of χ_D, χ_A, and ϵ where the excitation tunnels between the sites with the optimal efficiency. In this exceptional resonant point, the complex nonlinear oscillation becomes very simple and very fast. Below the Jupyter Notebook code for the nonlinear dimer is displayed that finds numerically the time evolution for the TET resonance condition.

```
# Import necessary libraries
import numpy as np
import matplotlib.pyplot as plt
from scipy.integrate import solve_ivp

# Update plot settings for better visualization
```

```python
plt.rcParams.update({'font.size': 14})

# Function defining the differential equations
#   for the nonlinear dimer (Discrete Nonlinear Schr\"{o}dinger equation)
def dnls_dimer(t, y, V, e_d, e_a, chi_d, chi_a):
    # psi_d: wavefunction at site D
    # psi_a: wavefunction at site A
    psi_d, psi_a = y

    # Calculate the time derivatives of psi_d and psi_a
    dt_psi_d = -1j * (e_d * psi_d + V * psi_a - chi_d * abs(psi_d)**2 *
        psi_d)
    dt_psi_a = -1j * (e_a * psi_a + V * psi_d - chi_a * abs(psi_a)**2 *
        psi_a)

    # Return the derivatives as a single array
    return np.append(dt_psi_d, dt_psi_a)

# Parameters for the dimer system
V = 1    # Coupling strength between the sites
t_span = [0, 10]   # Time interval for integration
y0 = np.array([1 + 0.j, 0.0 + 0.j])  # Initial conditions
 #   for psi_d and psi_a (complex numbers)

# Energy levels and nonlinear interaction strengths for sites D and A
e_d, e_a = -4.25, 0.0   # On-site energies
chi_d, chi_a = -4.25, 4.25   # Nonlinear interaction coefficients
 #   (TET condition)

# Solve the system of ODEs using the Runge-Kutta 45 method
sol = solve_ivp(dnls_dimer, t_span, y0, method='RK45',
            args=(V, e_d, e_a, chi_d, chi_a), max_step=0.01)

# Plotting the results
fig = plt.figure(figsize=(12, 8))

# Plot the squared modulus (occupation probability) of psi_d
 #   and psi_a as a function of time
plt.plot(sol.t / V, abs(sol.y[0])**2, label=r"$\psi_D$")  # Occupation
 #   probability of site D
plt.plot(sol.t / V, abs(sol.y[1])**2, label=r"$\psi_A$")  # Occupation
 #   probability of site A

# Label the axes and add a grid and legend
plt.xlabel("Time t (1/V)")
plt.ylabel("Occupation probability")
plt.grid()
plt.legend()

# Display the plot
plt.show()
```

9.1 The DNLS Donor-Acceptor Dimer 135

In the specific dimer case, it was possible to find the nonlinear resonance through analytical means, but this is not generally true. Furthermore, there could be regions in the parameter space where other resonances appear even not as pronounced as the specific one. The application of machine learning methods can assist in discovering more general resonances especially in cases that are not integrable.

9.1.2 Machine Learning with Physics Methodology

We saw that the TET model is analytically solvable for the dimer, but it is very difficult to address larger systems with the same methodology. We now implement a solution based on a machine learning approach that can be then generalized further to cover more extended systems [3]. We follow an approach similar to the one exposed in Chap. 8 for the nonlinear dimer. The methodology blends machine learning with the physics of the problem and utilizes the numerical solution of the latter as input data for the machine learning model [4]. We start by first solving numerically the linear system, i.e., $\chi_D = \chi_A \equiv 0$. We use the 4th order Runge-Kutta method with an integration step $= 0.001$ and set as maximum integration time three times the system characteristic time, T_c, defined as $1/V$. Subsequently, we evaluate the time the linear system attains a maximum in the probability at the acceptor site p_A, i.e., $T_{linear}^{max(p_A)}$. We designate this time as $T_{max} \equiv T_{linear}^{max(P_A)}$. The time T_{max} is important since it determines the range of training in the machine learning part. We know from the nonlinear dimer analysis that nonlinearity generally reduces the transfer period of complete motion to the other side, and thus the time based on the linear case is longer than the nonlinear cases of interest. Using the physics intuition for the problem is significant for the application of the AI-related methods.

We create a physics motivated machine learning model using Keras as implemented in TensorFlow 2.4. The two trainable variables are the nonlinearity parameters χ_D and χ_A, while we fix the value of the nondegeneracy ϵ. The machine learning model then needs to discover the nonlinear resonance values of Eq. (9.13). The variables χ_D and χ_A are initialized completely randomly. It is important to define a loss function $\mathcal{L}$ that is motivated by the physics of the model and that leads to the desired result. We select the following function:

$$\mathcal{L}(T_{max}, \chi_D, \chi_A) = 1 - |\psi_A(T_{max}, \chi_D, \chi_A)|^2, \quad (9.14)$$

where $|\psi_A(T_{max}, \chi_D \chi_A)|^2$ is the probability of the system being at the target state, A, at time, T_{max}, for specific values of χ_D and χ_A. This definition ensures that the loss function is expressed in terms of the variables χ_D, and χ_A and that the optimal χ_D and χ_A values will maximize the transition probability at time T_{max}.

In order to optimize the trainable parameters χ_D and χ_A, we use a loop and identify the values of the unknown values of χ_D and χ_A that lead to the optimal loss function. In each epoch of the training loop, we may integrate

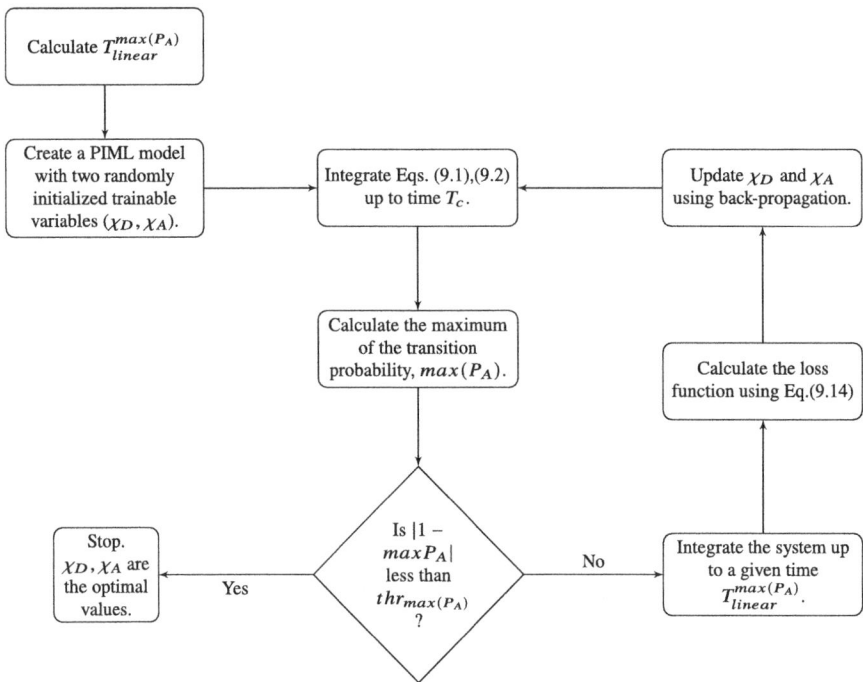

Fig. 9.1 Flowchart with the structure of the AI algorithm used in the search for the optimal nonlinear resonant configuration

Eqs. (9.10), (9.11), (9.12) up to a certain time T_c that is larger than T_{max} and find the maximum probability $max(P_A)$ in this time range. From this procedure we calculate the loss function using Eq. (9.14) in the time range up to T_{max}. Subsequently, we evaluate the gradients of Eq. (9.14) with respect to the trainable variables χ_D and χ_A and update their values using backpropagation with an adaptive learning rate. We start with a learning rate equal to one and reduce it by 5% when the gradients change sign. A change in the sign of the gradients indicates that we have passed the minimum, and thus we must now reduce the learning rate in order to avoid missing this minimal value. The training then stops when $|1 - max(P_A)|$ is less than a threshold value, $thr_{max(P_A)}$ or when the absolute value of the mean value of the twenty last values of the trainable parameters minus the mean value of the last two values of the trainable parameters is less than a threshold value, $thr_{\chi_{D/A}}$. We used the values $thr_{max(P_A)} = 10^{-8}$ and $thr_{\chi_{D/A}} = 10^{-4}$. The details of the procedures followed in the method used are shown in the flowchart (Fig. 9.1).

9.1.3 Results

The backpropagation procedure we introduce is able to optimize the dynamics based on the criterion set; the outcome is an optimal system with values $\chi_D = -\chi_A \equiv \epsilon$. The method we use was able to recover the condition that is used for targeted transfer systems (TET) [1]. In order to explore the range and convergence to the values that give the optimal transfer, we perform a grid search where we start with different initial conditions. In Fig. 9.2 we find the optimal TET values for a system with $\epsilon = 4.25$ and for ten randomly assigned initial conditions (blue filled circles). In all the cases the optimization concludes to $\chi_D = -\chi_A = 4.25$ (yellow star).

In Fig. 9.3 we show training details of one of the randomly assigned initial values of χ_D and χ_A. In the upper left graph we see the convergence of χ_D (red solid line) and χ_A (blue solid line) to the optimal values as a function of the number of epochs in the training loop. In the upper right, the loss function (black line) and the maximum transition probability, $max(P_A)$, (green line) as a function of the number of epochs in the training loop. The trajectory of the χ_D and χ_A variables during the optimization process is shown in the lower left graph and the transition probability for the optimal χ_D and χ_A values as a function of time, t versus the transition probability of the linear system in the lower right graph.

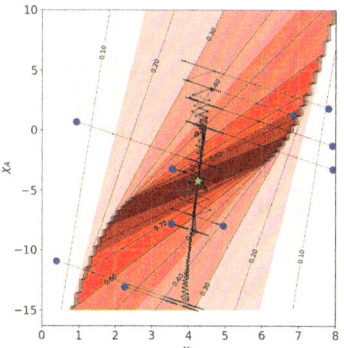
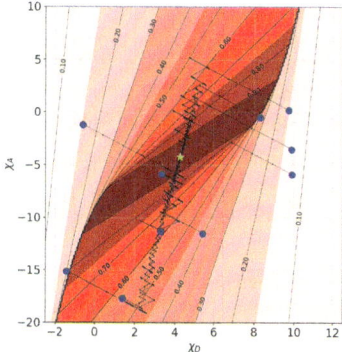

Fig. 9.2 The transition probability to the acceptor p_A landscape of a nonlinear donor-acceptor system as a function of χ_D and χ_A. Light red areas indicate low transition probability, and dark red areas high transition probability. We see ten randomly assigned initial values for χ_D and χ_A (blue filled circles) with their trajectories (black dashed dotted lines) and the optimal parameters for each one of the trajectories (yellow star). All the trajectories conclude to the same optimal values for χ_D and χ_A. Left: $\epsilon = 4.25$, V = 0.5. Right: $\epsilon = 4.25$, V = 1

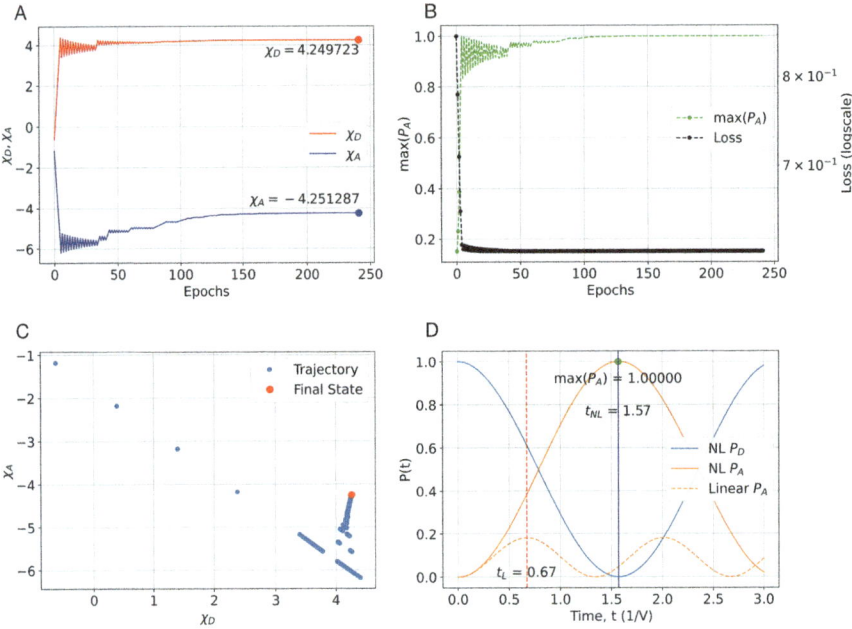

Fig. 9.3 (a) Convergence of χ_D (red solid line) and χ_A (blue solid line) to the optimal values as a function of the number of epochs in the training loop. (b) The loss function (black line) and the maximum transition probability (green line) as a function of the number of epochs in the training loop. (c) The trajectory of the χ_D and χ_A variables during the optimization process. (d) The occupation probabilities $P(t)$ in the donor (blue) and acceptor (red) sites for the optimal χ_D and χ_A values as a function of time, t

9.2 Transition over a Barrier into a Third State

We saw previously how effective is the machine learning approach in deriving the resonant TET condition for the analytically known case of the dimer. The analytical generalization of TET in a chain is very involved and not readily available. However, we now have the machine learning approach to our arsenal that can be used for these cases. A more complex situation is that where between the donor and acceptor units there is an intermediate state that for simplicity is a completely linear site as in Fig. 9.4. A situation similar to this occurs when the donor-acceptor pair is embedded in a lattice, and there is an intermediate state between them. In the photonic equivalent case this system corresponds to a linear fiber being placed in between the two nonlinear ones.

With the additional single intermediate site in the donor-acceptor system, the nonlinear transfer resonance becomes more elusive since the additional state creates a barrier that the excitation much overcome. The equations of motion for the Donor-Barrier-Acceptor system are:

9.2 Transition over a Barrier into a Third State

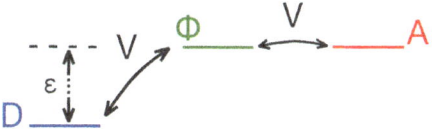

Fig. 9.4 Nonlinear trimer in site representation with the corresponding energy levels. The intermediate site labeled as Φ is a purely linear site with energy identical to that of the acceptor molecule

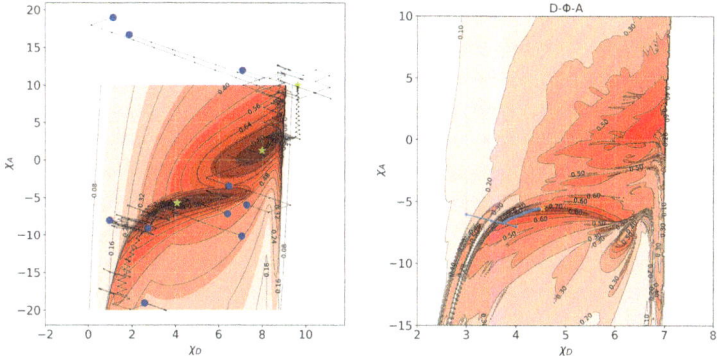

Fig. 9.5 Left: Donor-Barrier-Acceptor landscape with several randomly initialized (blue filled circles) trajectories (black dashed dotted lines) of the χ_D and χ_A variables during the optimization process. The star points to the position for the optimal values of χ_D and χ_A that lead to the maximum transfer. We note the presence of two optimal points, the main one and a secondary one on the upper right-hand side of the figure. Right: A zoom into the main optimal point

$$\begin{aligned} i\dot{\psi}_D &= -\epsilon\psi_D + V\phi + \chi_D|\psi_D|^2\psi_D \\ i\dot{\phi} &= \phantom{-\epsilon\psi_D +{}} V(\psi_D + \psi_A) \\ i\dot{\psi}_A &= \phantom{-\epsilon\psi_D +{}} V\phi + \chi_A|\psi_A|^2\psi_A, \end{aligned} \quad (9.15)$$

where we label with ϕ the probability amplitude for the intermediate "barrier" state.

We apply once again the optimization procedure detailed previously in analyzing the basic question, i.e., given a parameterization of the transfer rate V and the energy nondegeneracy ϵ what are the optimal donor-acceptor nonlinear values that lead to the most efficient transfer. The trimer dynamical system is in general chaotic, and thus we need to search in the complex space that includes chaotic systems for the class that has the desired optimal properties. Clearly, the transfer from donor to acceptor cannot be as efficient now compared to the dimer case.

In the case of the Donor-Acceptor system, all the trajectories converge to the theoretical values of χ_D and χ_A, i.e., $\chi_D = -\chi_A = \epsilon$, where $\epsilon = 4.25$. In the case of the Donor-Barrier-Acceptor system, the transition probability landscape is more complex containing not only a global maximum but also secondary local maxima (Fig. 9.5). In order to count the dynamics correctly, we increase the maximum

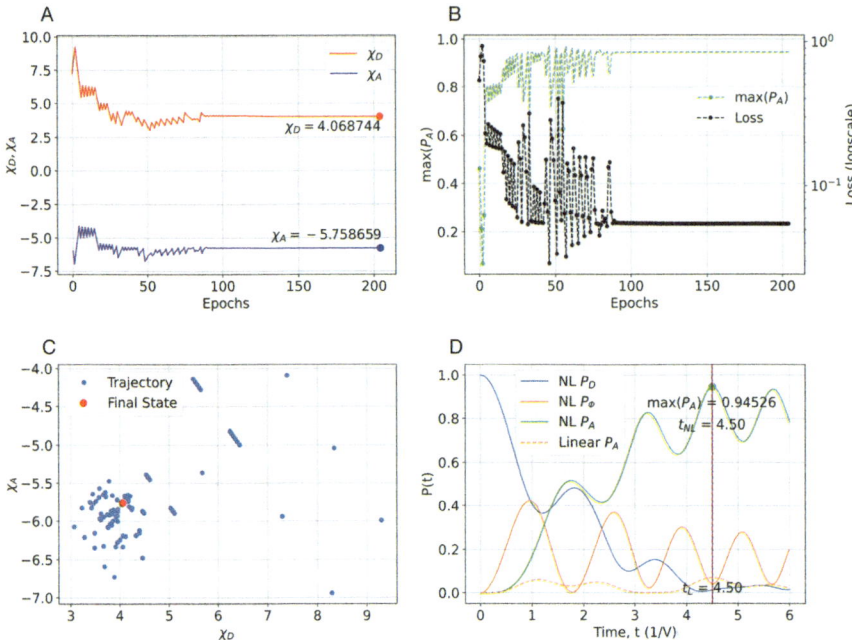

Fig. 9.6 (a) Convergence of χ_D and χ_A to the optimal values as a function of the number of epochs in the training loop. χ_D in solid red and χ_D in solid blue lines. (b) The loss function (black line) and the maximum transition probability (green line) as a function of the number of epochs in the training loop. (c) The trajectory of the χ_D and χ_A variables during the optimization process. (d) The occupation probability $P(t)$ at the donor (red) and acceptor (blue) sites for the optimal χ_D and χ_A values as a function of time, t

integration time up to six times the characteristic time, $1/V$, of the system. This ensures that almost all the trajectories will conclude to the global or at least to a local maximum of the transition probability. In Fig. 9.6 the details of the optimization process of one of the trajectories that concluded to the global maximum of the transition probability are presented.

Choosing an adaptive learning rate helps the training process, and the system converges to the parameters of the maximum transition probability.

9.3 Conclusions

In this chapter we introduced a special DNLS system that has both energy nondegeneracy but also different nonlinearity values. This more general case of the DNLS equation was specialized to the case of a dimer system with given nondegeneracy and two independent nonlinearity values, one for each site. We called this system a donor-acceptor dimer since it is inspired by special chemical reactions in photosynthetic and other systems [5, 6]. For this class of donor-acceptor systems, we know that there is an optimal set of parameters that make the unit fully resonant; this is the TET dimer. In order to find the optimal dimer through machine learning-type methods, we searched in the available parameter space. We used a form of backpropagation that enables efficient search in the space of systems in a way similar to actions in neural networks. The backpropagation procedure converges fast and recaptures the TET condition known analytically. This is quite significant since it shows that proper use of physics motivated machine learning may lead to discoveries that have solid mathematical basis. This also shows that proper application of AI-type methods has the capability to establish results and properties of complex dynamical systems.

The method was tested in the analytically known case and subsequently applied in a more complicated case where the results were not known. The latter is a donor-acceptor system with an intermediate site in between that plays the role of a barrier to the tunneling processes. It is not easy to find neither analytically or even numerically the nonlinear resonances that lead to optimal transfer in this system. The machine learning-motivated method worked well however in this more complex problem and found the desired parameter regimes. We note that the method is general and can also be applied in engineering applications where similar TET-type nonlinear resonances play a significant role [7].

The method introduced and applied in small DNLS-type systems can be used in order to construct nonlinear dynamical systems with prescribed properties without having to search explicitly complex resonance structures in these systems. Additionally, since the DNLS equation is a semiclassical equation that describes quantum mechanical transfer, one may also apply the method to strongly interacting quantum systems and design photonic as well as quantum metamaterials with specific properties.

9.4 Summary

- The targeted energy transfer dimer model consists of two sites where completely resonant transfer can take place.
- The TET model is based on the DNLS equation and in the simplest dimer case is fully integrable.
- The TET resonant condition can be recovered through machine learning methodology with a proper definition of the loss function.
- The method easily extends to more complex cases that cannot be handled as easily analytically.

References

1. G. Kopidakis, S. Aubry, G.P. Tsironis, Targeted energy transfer through discrete breathers in nonlinear systems. Phys. Rev. Lett. **87**, 165501 (2001)
2. A. Aubry, G. Kopidakis, A.M. Morgante, G.P. Tsironis, Analytical conditions for targeted energy transfer. Phys. B **296**, 222 (2001)
3. M. Raissi, P. Perdikaris, G.E. Karniadakis, Physics-informed neural networks: A deep learning framework for solving forward and inverse problems involving nonlinear partial differential equations. J. Comput. Phys. **378**, 686–707 (2019)
4. G.D. Barmparis, G.P. Tsironis, Discovering nonlinear resonances through physics informed machine learning. J. Opt. Soc. Am. B, **38**, C120 (2021)
5. S. Aubry, G. Kopidakis, A nonadiabatic theory for ultrafast catalytic electron transfer: a model for the photosynthetic reaction center. J. Biol. Phys. **31**, 375–402 (2005)
6. N. Almazova, S. Aubry, G.P. Tsironis, Targeted energy transfer dynamics and chemical reactions. Entropy **26**, 753 (2024)
7. A.F. Vakakis, O.V. Gendelman, L.A. Bergman, D.M. McFarland, G. Kerschen, Y.S. Lee, *Nonlinear Targeted Energy Transfer in Mechanical and Structural Systems* (Springer, 2009)

Part IV
Chaos and Spatiotemporal Complexity

Chapter 10
Dynamical Embedding with Autoencoders

Dimensionality of Chaotic Series

Abstract The quantitative analysis of complex systems may be assisted by the use of methods derived from neural networks. The analysis of chaotic time series presents an applied challenge toward determining the features involved and particularly the effective complexity they include. This chapter shows that the use of chaotic autoencoders made of neural networks leads to essential information on the time series and the chaotic systems they arise from. The autoencoders we construct map identically the input chaotic time series to the output. The process passes through the latent space that acts as an information bottleneck for this reconstruction. We estimate the dimension of the latent space numerically and find it smaller than possible embedding dimensions of the reconstruction. The dimension of the latent space hints toward the complexity of the dynamical system underlying the time series. Furthermore, we show that the constructed chaotic autoencoders produce maximal Lyapunov exponents that are very close to those obtained directly from the equations of motion of the chaotic systems. These features show that the constructed chaotic autoencoders are faithful representations of the time series generating dynamical systems, and through their latent space dimension, we obtain an estimate of their complexity.

10.1 Introduction

Nonlinear dynamical systems, either continuous or discrete, may involve chaos, i.e., irregular evolution similar to random, and stochastic motion. Continuous autonomous chaotic systems are described by at least three coupled ordinary nonlinear differential equations that when solved numerically result in trajectories with highly irregular features. In non-Hamiltonian systems, on the other hand, we typically find the trajectory to fall on a strange attractor, i.e., a distinct phase space structure with interesting, nontrivial features. Understanding the complex dynamics in phase space is significant especially in the inverse problem where given a time series of discrete data and need to assess whether it derives from a chaotic dynamical system or it is purely stochastic [1]. In the former case, we are interested in reconstructing the phase space of the associated dynamical system from the data [1].

While the methods developed over the last 30 or more years work quite well, they are many times cumbersome and involve additionally empirical inference. It would be thus very profitable to apply new methods based on artificial intelligence and explore their power and validity [2–5]. This is the aim of the present chapter where we apply unsupervised learning methods and attempt to reconstruct the dynamical features of the data without relying on the traditional methods.

In this application we will use autoencoders, a machine learning tool that is used widely as a dimensionality reduction technique. Autoencoders are neural network-based "unit projection operators" where the output is as close as possible to the input. In the process of this unit projection, data pass through the latent space where they suffer dimensionality reduction. The effective dimension of the latent space that works as information bottleneck determines the degree of possible data compression and thus eliminates original dimension redundancy. Since the neural network uses fewer nodes in latent space than the number of input nodes, we accomplish compression of the original data. This packing allows the detection of the basic information contained in the data. In order to effect compression, one introduces regularization of the data in the latent space using as parameter the dimensionality of the space. This is the approach that leads to the optimal number of operating nodes necessary for the original data reconstruction. This strategy allows us to find the essential number of nodes for the trained network that recapture the complexity of the original time series.

Once the minimal configuration in latent space is found, we need to compare the chaotic autoencoder with the original system. This is accomplished through the comparison of the corresponding maximal Lyapunov exponents, i.e., that of the original trajectories of the system with the ones calculated from the reconstructed trajectories generated through the autoencoder. The Lyapunov exponents are a general attribute of nonlinear dynamical systems as they allow to distinguish the local stability and invariant sets in a particular direction of the attractors. The largest Lyapunov exponent detects the mean divergence of two initially nearby trajectories. The positive value of the exponent defines the degree of chaos in a particular system [1].

An autoencoder is a neural network that maps the input data to the output [2]. It presents a method of unsupervised learning with the compression of data occurring in the latent space that forms a kind of information bottleneck, due to the smaller number of nodes it includes. The target data for an autoencoder are the input data. Traditionally, autoencoders are used as dimensionality reduction techniques and salient features of learning. The construction of an autoencoder includes two main parts. The first part is an encoder function, $h(x)$, that compresses the input information, x, while the second part is the decoder, $g(h)$ takes as input the compressed information, and by decompressing it reconstructs the initial input of the encoder. The output function can be written as $\hat{x} = g(h(x))$, where the input sequence should be learned "perfectly." In other words, the autoencoder attempts to copy the data to the output, while first they pass through a compression bottleneck. The learning results are obtained through a minimization of the appropriate loss function.

In this chapter we focus on the construction of "chaotic autoencoders," i.e., autoencoders that may analyze chaotic systems [5]. In the next section, we detail features of dynamical systems such as phase space reconstruction and the largest Lyapunov exponents and report on the methods for their calculation. Subsequently, we define the chaotic autoencoders for three prototypical dynamical systems and show how the information compression operates. We then show how the Lyapunov exponents are evaluated in the chaotic autoencoders.

10.2 Description of the Chaotic Systems

A deterministic chaotic signal behaves as something intermediate between a regular and a stochastic time series. The nonlinear aspect of this behavior of the signal makes difficult the prediction and classification of the physical system, from which the motion comes from [6]. A chaotic motion produces specific structures in phase space. Systems of differential equations with three or more degrees of freedom may show chaos in their evolution in real or phase space. The number of degrees of freedom of a system describes the number of the necessary independent variables to specify the dynamical state of the system. Chaotic systems are described through their time evolution equations, the values of their parameters, and the selected initial conditions. We may express the ordinary differential equations guiding the evolution of a state using a variable $x(t) = [x_1(t), x_2(t), \ldots, x_i(t)]$, where $x_i(t)$ is the variable describing the i-th degree of freedom at time, t, as

$$\frac{dx(t)}{dt} = f(x(t)), \tag{10.1}$$

where f describes the nonlinear function that characterizes the equations of motion of the system. Given the system of differential equations guiding a dynamical system, we can solve them numerically and follow the evolution in time. Many dynamical systems have been investigated using this methodology. Here, we focus on prototypical systems well-known in the literature of chaotic dynamics, viz. the Rössler system and the Lorenz system. Since we want to have a more general idea of the effectiveness of our method, we use both the Lorenz-63 system that contains three degrees of freedom and the Lorenz-96 one with multiple dependent variables and external forcing. The differential equations for the three nonlinear systems and the parameter regimes we use in the analysis follow:

Lorenz System
The classical Lorenz system is the prototypical chaotic system discovered by Lorenz in 1963. It consists of three nonlinear, coupled differential equations that represent variables x, y, z related to meteorological quantities. We will refer to this model as Lorenz-63:

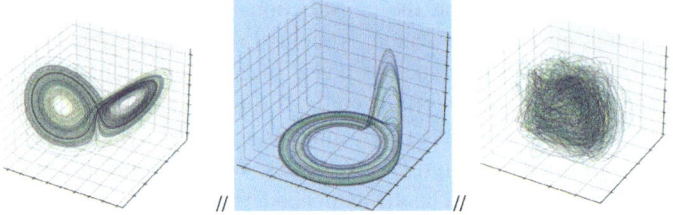

Fig. 10.1 Strange attractors for three chaotic systems. Left: Lorenz, middle: Rössler, and right: Lorenz-96

$$\frac{dx}{dt} = \sigma(y - z) \tag{10.2}$$

$$\frac{dy}{dt} = x(\rho - z) - y \tag{10.3}$$

$$\frac{dz}{dt} = xy - \beta z. \tag{10.4}$$

In the model we also have three parameters σ, ρ, and β; depending on the actual values of these parameters, the dynamics is different. In Fig. 10.1 we show the phase space behavior of the Lorenz model for the values of parameters $\sigma = 10$, $\rho = 28$, and $\beta = 8/3$. We notice a beautiful double ring structure in phase space; this is the Lorenz attractor. It is a *strange attractor*, i.e., an invariant set where most trajectories in this parameter regime lie. The surface of the attractor has fractal dimension [6].

Rössler System

This is a relatively simpler dynamical system described by the following set of equations:

$$\frac{dx}{dt} = y - z \tag{10.5}$$

$$\frac{dy}{dt} = x + ay \tag{10.6}$$

$$\frac{dz}{dt} = b + z(x - c), \tag{10.7}$$

where x, y, z are the three dynamical variables while a, b, c the free parameters of the model. The chaotic attractor seen in Fig. 10.1 is in the parameter regime $a = 0.1$, $b = 0.1$, and $c = 14$. A strange attractor is an invariant set on which most trajectories in this parameter regime fall and has a fractal dimension.

Lorenz-96 System

The Lorenz-96 model is a higher dimensional system that may be seen as a lattice model; the equation for the i-th dependent variable $x_i(t) \equiv x_i$ is

$$\frac{dx_i}{dt} = (x_{i+1} - x_{i-2})x_{i-1} + F, \tag{10.8}$$

where F is a constant parameter and N the number of degrees of freedom. We may view this set as a nonlinearly coupled chain of overdamped oscillators driven by the constant force F. In Fig. 10.1 we show a three dimensional projection of the attractor for $N = 40$ and $F = 8.15$.

10.2.1 Delay-Coordinate Embedding

A physical system can only be measured discretely in time, with a given time step, and thus it can be represented as a time series of data. In the case of a chaotic system, we can use several methods to determine the features of the chaotic data. One of these methods is the time-delay embedding or reconstruction scheme [1]. This method allows for the recovery of the entire multidimensional observation in state space based on the time series of just one of its variables. Let us assume that a time series, X, represents a trajectory of a physical system recorded with a fixed time step as an N-point time series. The reconstructed trajectory, X, can be represented as an $(M \times m)$ matrix

$$X = \begin{bmatrix} X_1, X_2, \ldots, X_M \end{bmatrix}^T, \tag{10.9}$$

where $M = N - (m-1)\tau$, m is an embedding dimension, τ is a time lag (delay), and X_i is a vector representing the state of the system at discrete time i, given by

$$X_i = \begin{bmatrix} x_i, x_{i+\tau}, \ldots, x_{i+(m-1)\tau} \end{bmatrix}, \tag{10.10}$$

with x_i being the i-th data point of the time series. In order to find the embedding dimension for a dynamical system, one can implement the Takens theorem [1]. Takens introduced the idea that a complete part of dynamics with basic dimension d_s might be observed in an embedding space if the latter has $2d_s + 1$ or higher dimensions. There are a number of practical implementations of Takens theorem although these methods are not necessarily universal. In the present exposition, we use a time lag equal to 1 and an appropriate embedding dimension that satisfies the Takens theorem.

10.3 Autoencoder Methodology

We focus now on preparing the data and setting up the autoencoder for the chaotic systems. This algorithm is general and can be applied in chaotic time-series analysis or other time series with seemingly noisy structure.

10.3.1 Data Preparation

We use data from the three chaotic systems referred to previously, viz. that of Rössler, the classical Lorenz system or Lorenz-63 and the Lorenz-96 model. In order to generate data we integrate the differential equations of each system using the fourth order Runge-Kutta method, with a fixed time step equal to 5×10^{-3} for a total of 3×10^5 time steps. We remove the first 5×10^4 time steps that are considered transient. In order to train the autoencoder we use only the x-coordinate of the Rössler and the Lorenz-63 systems and the 19th coordinate (out of 40) of the Lorenz-96. The available 2.5×10^5 time steps of each training sequence are normalized to a [0, 1] range and split into two distinct sets; 80% of the data form the training set while the rest 20% the test set. Both training and test sets are further split into smaller segments using a sliding window of size W, i.e., determined from the size of the input window.

10.3.2 Chaotic Autoencoder Construction

We construct a unique stacked autoencoder for each chaotic system composed of two symmetric networks, the encoder, $h(x)$, and the decoder, $g(h)$, that will be used to data reconstruction. The input layer has a number of nodes equal to the window size W used to split the input sequence; in the initial application it is taken to be $W = 30$. Two fully connected hidden layers constitute the *encoder layer* and follow the input layer. They are constituted with 22 and 15 nodes, respectively, with a sigmoid activation function for the former and a ReLU one for the latter. The *latent space* layer follows this initial pair of hidden layers. It consists of a fully connected layer of ten nodes with a ReLU activation function. The *decoder layer* is simply a mirror image of the encoder structure.

The critical junction in the autoencoder is the latent space layer. While it has a fixed number of nodes, not all of them are necessary for processing the input information. In order to find the minimal number of nodes that are necessary, we apply L_1 regularization, or Lasso Regression, in latent space—this has the effect of practically eliminating nodes with limited use or information. The L_1 regularization is applied to the output layer as an additional term in the overall loss function and depends on the regularization parameter α that is a hyperparameter of the model. The role then of the regularization in latent space is to optimize the actual *number* of nodes necessary to recapture the input sequence information. The larger the regularization parameter α, the more we constrain the autoencoder, and thus we increase the error in the reconstructed sequence. The smaller the regularization parameter, the larger the active part of the latent space, and thus we decrease the error in the reconstructed sequence.

After the end of the training process, we may perform a statistical analysis of the relative contributions of the number of active nodes in the latent space that determine

10.3 Autoencoder Methodology

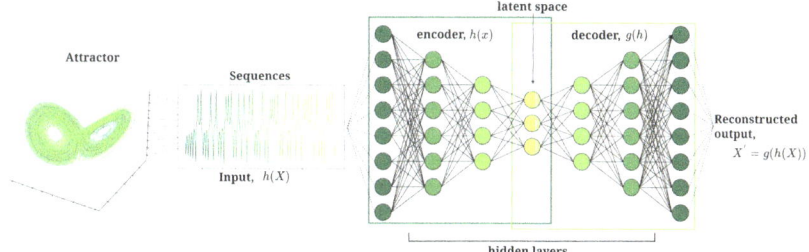

Fig. 10.2 A schematic representation of the chaotic autoencoder model. One variable, for instance x, that is a one dimensional projection of the system trajectory, is fed into the encoder model, $h(x)$, suffers compression in the latent space, and then follows through the decoder, $g(h)$. The combined action of the encoder-latent space-decoder is to reconstruct optimally the original information, i.e., $\hat{x} = g(h(x))$

the accuracy of the reconstructed sequence in the output. The regularization parameter separates the nodes into the ones with non-important contribution to the reconstruction of the sequence and the ones with significant contribution. The average number of the latter nodes in latent space with significant contribution to the reconstruction is the basic contribution of the latent space regularization. This is the essential information that is linked to the minimal information necessary to reconstruct the chaotic attractor and thus may be connected to the embedding space dimension.

The autoencoder model is trained for 7500 epochs with a batch size equal to 32, using the Adam stochastic optimization method with a learning rate of 0.001, while all its weights are initialized using the Xavier uniform initializer as implemented in TensorFlow/Keras. In Fig. 10.2 we present a schematic representation of the chaotic autoencoder model.

We have already seen that an autoencoder is a neural network that acts as an identity operator and maps the input data ideally to identical output data. In order to perform this operation through the latent space, the autoencoder must learn by minimizing a loss function in the output. The simplest example of a loss function is that of the mean squared error (MSE), i.e.,

$$MSE(x, \hat{x}) = \frac{1}{W} \sum_{k=1}^{W} \left(x_k - \hat{x}_k \right)^2, \tag{10.11}$$

where W is the length of the input sequence, x represents the real input data (a vector of x_k real samples), $\hat{x} = g(h(x))$ represents the reconstructed output data (a set of $\hat{x}_k$ reconstructed samples) of a general machine learning system, while the loss function "punishes" the difference between x and $\hat{x}$, i.e., the real and reconstructed data. In order to train the autoencoder we apply additionally the sparsity penalty, i.e., the L_1 regularization, $\Omega(h) = \alpha |h(x)|$, on the output of the encoder, $h(x)$, after the

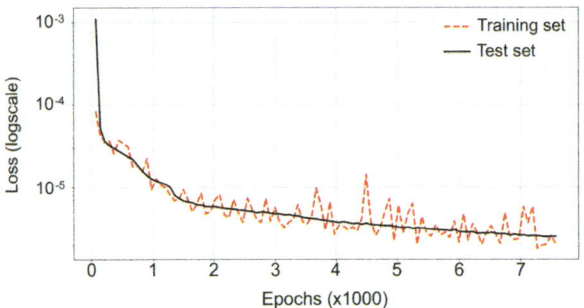

Fig. 10.3 Lorenz-63 model: evolution of training (red dashed line) and test (solid black line) set loss function (logarithmic scale) as a function of the number of epochs of training

activation function, where α is the regularization parameter. The total loss function, L, is then

$$L(x, \hat{x}) = MSE + \Omega(h(x)). \qquad (10.12)$$

The addition of the L_1 term in L enables the autoencoder to limit the number of nodes in latent space to a minimum necessary for the optimal reconstruction of the original signal. This type of regularization forces the participation of nodes in latent space to be zero when their contribution to the output of the latent space is less than the regularization parameter. Since we are interested to investigate the performance of the model as a function of the regularization parameter α and find the minimum number of nodes that can correctly reconstruct the input sequences, we may fix all the other parameters of our model and train it only once. We can thus compare the results for the different values of the regularization parameter. In Fig. 10.3, we present the evolution of the loss function during the training process for the training and test sets of the Lorenz-63 system.

10.3.3 Latent Space Dimension

The regularization parameter, α, controls the size of the latent space and plays a crucial role in determining the number of the important nodes in the latent space. In this approach, we consider important the nodes in the latent space that output a value larger than the regularization parameter, α. Thus, we calculate the average of nonzero nodes in the test set as a function of α. The results for the three systems under consideration are shown in Fig. 10.4.

10.3 Autoencoder Methodology

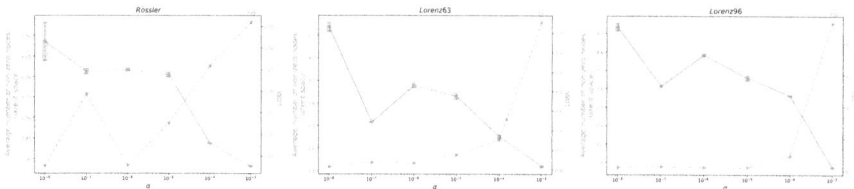

Fig. 10.4 The average number of nonzero nodes at the latent space (navy blue solid line), including the standard deviation of it (error bars), and the corresponding value of the loss function (red dashed line) of the test data as a function of the L_1 regularization parameter, α, for the three chaotic systems we study

10.3.4 Largest Lyapunov Exponent as a Metric

The Lyapunov exponents are an essential concept of nonlinear dynamics and present specific features of chaotic systems. Two trajectories of the attractor initiated at points displaced infinitesimally from each other will diverge exponentially if the attractor is chaotic. The exponential deviation is described by the largest Lyapunov exponent (LLE). A spectrum of Lyapunov exponents can also be defined [1]. The LLE of a system can be calculated using two approaches. The first approach uses a Jacobian matrix and can be applied when the mathematical model describing the evolution of the system is known. This method gives the Lyapunov spectrum. The second one can be used to estimate the LLE of a time series by applying the definition of the local divergence on a segment of a given trajectory. Estimating the LLE from a time series faces several numerical difficulties, such as having to deal with the noise in the data. The basic definition of the LLE, λ_1, can be written as

$$d(t) = Ce^{\lambda_1 t}, \tag{10.13}$$

where $d(t)$ is an average trajectory distance at time t and C is a constant (usually normalized by the initial separation of neighbors). We divide the time, t, into steps of Δt and assume that at time $t = i \times \Delta t$, the distance of the jth pair of nearest neighbors, $d_j(i)$, diverges approximately at a rate given by the LLE [5]:

$$d_j(i) \approx C_j e^{\lambda_1 i \Delta t}, \tag{10.14}$$

where C_j is the initial distance of the selected pair of points. By taking the logarithm of Eq. (10.14) we have that

$$ln d_j(i) \approx ln C_j + \lambda_1 (i \Delta t). \tag{10.15}$$

Equation (10.15) represents a set of, M, (all the values of index j) approximately parallel lines that can be fitted using least squares to calculate an averaged value and the standard deviation of the LLE.

In order to have a clear picture on how close is the autoencoder to the original chaotic system, we compare the average value and the standard deviation of the LLE evaluated using the original and the reconstructed trajectories. The reconstructed dynamics obtained by the autoencoder depend on the parameter in L_1 regularization—α. We report these findings and the dependence of LLE on the embedding dimension window below. The important outcome of the autoencoder analysis is the determination of the effective minimum size of the latent space of the autoencoder. The assumption here is that the reconstruction of the attractor is determined in an embedding space, the dimension of which is determined directly by the latent space dimensionality. The size of the latent space depends on the regularization parameter α and the window size of input sequence of the time series, W. Once the autoencoder is trained, we determine the largest Lyapunov value it corresponds to. In other words, the construction of the autoencoder system behaves as a twin of the original system and shares with it basic properties, such as chaos, embedding dimension, Lyapunov exponent, etc. We may then use the autoencoder-based system copy in order to find properties of the original chaotic system.

10.3.5 Regularization Parameter Dependent Latent Space Size

When we apply the autoencoder with the additional L_1 regularization of the chaotic data in the way described previously for various values of the hyperparameter α, we obtain valuable information on the "degree of chaoticity" of each nonlinear system. In Fig. 10.4, we show the average number of nonzero nodes at the latent space, i.e., nodes with output value larger than α, as well as the loss of the reconstructed sequence, with respect to the input sequence, as a function of the L_1 regularization parameter, α. We observe that in all three chaotic systems the behavior is similar. At low values of α where the L_1 term is not very significant and thus does not constrain the latent space, the number of nodes is relatively large while the total loss relatively low. As the ridge regression becomes more important, the number of nodes reduces and for a certain range of α-values reaches a plateau. The loss in the regime has also a plateau-like behavior, modulo numerical error. In the final regime of large hyperparameter α, the loss increases rapidly since the L_1 term controls it and reduces the number of nodes in latent space further.

The intermediate α-regime where the loss is low is the appropriate regime where the information bottleneck of the latent space is balanced: It has an appropriate number of nodes so that the autoencoder faithfully reconstructs the chaotic system with reasonably low loss. We find for the Rössler system that *two nodes* may be optimal for this process, for the classical Lorenz system approximately *three nodes* while for the mode complex Lorenz-96 model closer to *four nodes*. We may associate these numbers as the effective dimensionality of the embedding space of the chaotic time series. The fractal dimension of the Rössler attractor is estimated to be slightly above 2, i.e., 2.01–2.02, the one of the Lorenz attractor 2.06, while the Lorenz-96 model has more complex attractor structure. We thus see that the

10.3 Autoencoder Methodology

Table 10.1 Projections of the largest Lyapunov exponent obtained from autoencoder trained models for different regularization parameters α. In parentheses is the calculated error, while "Input" refers to the Lyapunov exponent projection as calculated from the chaotic model. The projections are along the x-axis for Rössler and Lorenz-63 models and the 20-th site for Lorenz-96

		Rössler	Lorenz-63	Lorenz-96
Sequence	α	X	X	20
Input	–	1.73 (0.08)	1.17 (0.30)	3.03 (0.35)
Reconstructed	10^{-3}	1.76 (0.05)	1.16 (0.42)	2.63 (0.73)
	10^{-4}	1.74 (0.02)	1.14 (0.13)	2.79 (0.53)
	10^{-5}	1.77 (0.09)	1.14 (0.16)	2.83 (0.44)
	10^{-6}	1.78 (0.09)	1.14 (0.26)	3.01 (0.22)
	10^{-7}	1.76 (0.07)	1.15 (0.29)	2.83 (0.56)
	10^{-8}	1.73 (0.25)	1.17 (0.27)	3.04 (0.64)

hierarchy obtained through the latent space dimensionality is compatible to the complexity that each model pertains.

To further demonstrate the efficiency of the reconstructed trajectories by the autoencoder model, we calculate the largest Lyapunov exponent for the input and the reconstructed sequences of each one of the regularization parameter values (Table 10.1). We calculate the LLE by creating two trajectories of each attractor. We choose a starting point for one of them, and then we displace it by 10^{-7} for the second trajectory. We create the two trajectories by solving the equations of each attractor, and then we use the trained autoencoder of the corresponded system to reconstruct the trajectories. In Table 10.1, we present the projections of the largest Lyapunov exponents in the relevant coordinates for the input and the reconstructed sequence for various values of the regularization parameter, α. The results indicate that the reconstructed sequences created by models with α smaller than or equal to 10^{-4} have the same characteristics as the sequences obtained by solving the deterministic system of equations. The error in the LLE is less than $\pm$ 2%. A similar analysis is shown in Table 10.2 for different input data lengths W.

10.3.6 Latent Space Dimension as a Function of the Size of the Input Sequence W

It is important to know how the size of the information passed to the autoencoder (the size of input sequence) affects the performance of the model. In Fig. 10.5, we present the average number of nonzero nodes in the latent space that were used to reconstruct the input sequence as a function of the size of the input sequence, W. We fix the regularization parameter α to 10^{-5} and range the size of the input sequence, W, from 9–37 time steps using step size 7 time steps.

Table 10.2 Comparison of projections of the largest Lyapunov exponent obtained from autoencoder trained models for different input lengths W. In parentheses is the calculated standard deviation, while "Input" refers to the Lyapunov exponent projection as calculated from the chaotic model. The projections are along the x-axis for Rössler and Lorenz-63 models and the 20-th site for Lorenz-96

		Rössler	Lorenz-63	Lorenz-96
Sequence	W	X	X	20
Input	–	1.73 (0.08)	1.17 (0.30)	3.03 (0.35)
Reconstructed	9	1.77 (0.02)	1.15 (0.37)	2.92 (0.64)
	16	1.73 (0.21)	1.13 (0.18)	3.05 (0.57)
	23	1.73 (0.21)	1.14 (0.40)	2.92 (0.59)
	30	1.76 (0.14)	1.14 (0.29)	2.97 (0.76)
	37	1.74 (0.16)	1.14 (0.18)	3.03 (0.36)

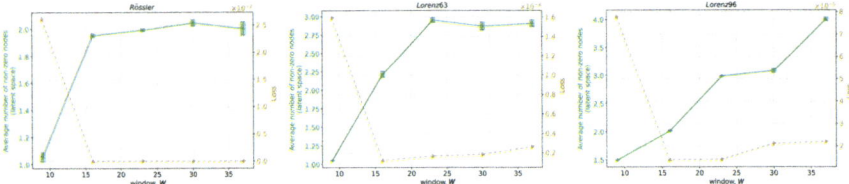

Fig. 10.5 The average number of nonzero nodes at the latent space (navy blue solid line), including the standard deviation of it (error bars), and the corresponding value of the loss function (red dashed line) of the test data as a function of the size of the input sequence, W, for the three systems

In the case of the Rössler system, the autoencoder needs two nonzero nodes in the latent space to reconstruct a sequence of more than 16 time steps. Three nonzero nodes are needed in the Lorenz-63 model to reconstruct a sequence of 23 time steps. We observe that the Lorenz-96 system is more complex than the other two, and the adequate number of nodes to represent it depends strongly on the desired accuracy. The error was less than 7% in all cases. An analysis of the Lyapunov exponents shows that the original and reconstructed time series give similar Lyapunov exponents.

10.4 Conclusion

In this chapter we connected chaotic dynamical systems with neural network-based autoencoders and found the minimal set of neural network nodes that may reproduce these dynamical systems. Autoencoders are deep neural networks with a specific mirror-like structure interrupted by a layer that operates as information bottleneck. In this application we use chaotic signals and tried to find the minimal dimensions of the latent space that reproduce faithfully the chaotic systems. In a sense we constructed a chaotic autoencoder twin system that shares the properties of the original, complex system. The dimensionality of the latent space is the crucial parameter that determines the minimal information necessary for a faithful reproduction of the chaotic system. We found that the dimensionality of latent space is relatively low but different for each chaotic system we investigated. Furthermore the average values obtained are compatible with the embedding attractor space dimensions and the fractal attractor dimension itself.

Intuitively one might think that the latent space is in some sense the minimal space required to represent a chaotic dynamical system. As a result, the latent space dimension should be smaller or at least equal to the embedding dimension since the latter gives the framework space for analyzing the chaotic series. We observe that this feature is true in the examples we addressed since we find latent space dimension smaller than or equal to the number of differential equations that generate respective data. An additional appealing outcome of the present analysis is that the largest Lyapunov exponent that is connected with the autoencoder system appears to be very close to the one determined directly from the dynamical system. Clearly, these two exponents are evaluated differently since the chaotic autoencoder cannot by construction generate its dynamics and depends on the input time series. The Lyapunov exponent in this case is obtained by observing the proximity of two nearby chaotic series with a very small difference in their initial conditions. However, the autoencoder can generate similar divergence in the difference of the two series as that of the original system shows that it provides a faithful representation of the original system. In other words, we find that the chaotic autoencoder provides a faithful representation of time-series segments and the more intricate internal dynamics of the chaotic system.

It is both intellectually appealing and practically important that we may represent a complex physical system with a deep neural network such as an autoencoder. In industry, the construction of a *digital twin* of a certain machinery or other setup is both very useful and financially efficient. One may study a real system digitally before it is actually constructed and experiment with different variations of it without having to fabricate it. The fact that even chaotic systems can be represented by special types of deep networks is very reassuring that the twinning processes are in principle reliable. Furthermore, due to the control we have on the latent space, the use of autoencoders appears very useful in extracting specific information for complex systems. It is possible that more sophisticated systems such as variational or probabilistic autoencoders will render more precise information on the complexity of nonlinear systems.

10.5 Summary

- Autoencoders reproduce chaotic systems.
- The latent space dimensionality is compatible to the embedding dimension and the fractal dimension of the chaotic attractor.
- Original and reconstructed systems share close values of the largest Lyapunov exponents.

References

1. H. Abarbanel, *Analysis of Observed Chaotic Data* (Springer, New York, 1996)
2. I. Goodfellow, Y. Bengio, A. Courville, Y. Bengio, *Deep Learning*, vol. 1 (MIT Press, 2016)
3. Z. Lu, J. Pathak, B. Hunt, M. Girvan, R. Brockett, E. Ott, Reservoir observers: model-free inference of unmeasured variables in chaotic systems. Chaos **27**, 041102 (2017)
4. P.R. Vlachas, W. Byeon, Z.Y. Wan, T.P. Sapsis, P. Koumoutsakos, Data-driven forecasting of high-dimensional chaotic systems with long short-term memory networks. Proc. R Soc. A. **474**, 2213 (2018)
5. N. Almazova, G.D. Barmparis, G.P. Tsironis, Analysis of chaotic dynamical systems with autoencoders. Chaos **31**, 103109 (2021)
6. E.N. Lorenz, The local structure of a chaotic attractor in four dimensions. Phys. D **13**, 90–104 (1984)

Chapter 11
Chimeras

AI for Complex Spatiotemporal Modes

Abstract Chimeras are complex spatiotemporal modes that feature chaos within order, i.e., they contain a chaotic part intermingled spatially with an ordered one. They appear in many strongly coupled nonlinear lattice systems usually with long-range interactions and are long lived. It is speculated that the brain of mammals operates in chimera states in several occasions. In this chapter we focus on chimeras in artificial systems such as metamaterials and use machine learning in order to predict their evolution. Since the lattices we focus on are rather complex, a simple data approach is not very successful. In this direction we use, in addition to the machine learning method, certain real time sensors we call "observers." These sensors provide ground-truth information in certain spatial locations of the lattice at all times—this aspect is very important for the forecast through machine learning. We utilize three machine learning methods, the simpler feed-forward neural network (FNN), the Long Short-Term Memory (LSTM) networks, and the reservoir computing (RC) recurrent neural networks. From the analysis we find that even a small number of observers greatly improve the data-driven model-free long-term forecasting capabilities of all methods.

11.1 Introduction

In this chapter we focus on the use of machine learning methods in complex processes in artificial or synthetic media called metamaterials. They are man-made systems constructed through repetition of specific units that generate new properties for the extended system. More specifically we will produce and use data for *SQUID metamaterials* that are a type of superconducting media [1–6]. The acronym SQUID stands for a superconducting quantum interference device that is a very unique nonlinear element based on superconductivity and the Josephson effect. The simplest version of a SQUID consists of a superconducting ring interrupted by a Josephson junction (Fig. 11.1a); this device is a highly nonlinear resonator with a strong response to applied magnetic fields. The SQUID metamaterials exhibit peculiar magnetic properties including negative diamagnetic permeability [2]. Applied alternating fields induce supercurrents in the SQUID rings, coupling

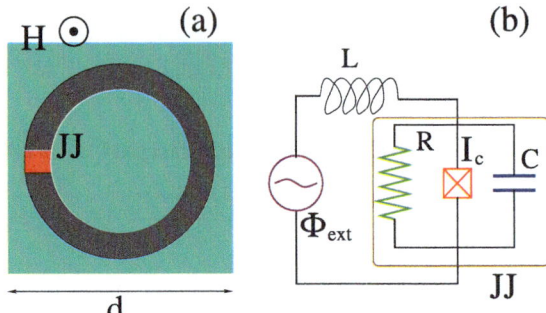

Fig. 11.1 Schematic of a SQUID unit in an alternating magnetic field (**a**) and equivalent electrical circuit (**b**). A SQUID metamaterial is formed by repetition in one or two dimensions of the unit in (**a**) with a certain lattice spacing d_0 [2]

them together through dipole-dipole magnetic forces; although weak due to its nature, the interaction results in nonlocal coupling between the SQUIDs, since it falls off as the inverse cube of their distance.

In order for the SQUID to produce a chimera state, it needs to be coupled with other similar units and form a type of superconducting metamaterial. The collective chimera states appear almost generically in networks of interacting nonlinear oscillators and have applications ranging from neurobiology to materials science [7, 8]. A *chimera state* is characterized by the coexistence of synchronous and asynchronous clusters of oscillators, even though the latter are coupled symmetrically and are identical. This remarkable collective behavior emerges also in SQUID metamaterials driven by an alternating magnetic field in the presence of weak dissipation. In order to form a model SQUID chimera, we need to introduce many SQUIDS that are interacting, initialize them with random fluxes, and follow numerically the dynamics. We briefly describe the mathematics of the problem below.

11.1.1 SQUID Metamaterials with Long Range Coupling

Let us consider a one dimensional linearly arranged array of N identical SQUIDs coupled together magnetically through dipole-dipole forces [9]. The magnetic flux Φ_n threading the n-th SQUID loop is

$$\Phi_n = \Phi_{ext} + L I_n + L \sum_{m \neq n} \lambda_{|m-n|} I_m, \quad (11.1)$$

where Φ_{ext} is the external flux in each SQUID, $\lambda_{|m-n|} = M_{|m-n|}/L$ is the dimensionless coupling coefficient between the SQUIDs at positions m and n, with $M_{|m-n|}$ being their corresponding mutual inductance, and

$$-I_n = C \frac{d^2 \Phi}{dt^2} + \frac{1}{R} \frac{d\Phi}{dt} + I_c \sin\left(2\pi \frac{\Phi_n}{\Phi_0}\right) \quad (11.2)$$

11.1 Introduction

is the current in each SQUID given by the resistively and capacitively shunted junction (RCSJ) model [1], with Φ_0 and I_c being the flux quantum and the critical current of the Josephson junctions, respectively. Within the RCSJ framework, R, C, and L are the resistance, capacitance, and self-inductance of the SQUIDs' equivalent circuit (Fig. 11.1b). Combination of Eqs. (11.1) and (11.2) gives

$$C\frac{d^2\Phi}{dt^2} + \frac{1}{R}\frac{d\Phi}{dt} + \frac{1}{L}\sum_{m=1}^{N}\left(\hat{\Lambda}^{-1}\right)_{nm}(\Phi_m - \Phi_{ext})$$

$$+ I_c \sin\left(2\pi\frac{\Phi_n}{\Phi_0}\right) = 0, \quad (11.3)$$

where $\hat{\Lambda}^{-1}$ is the inverse of the $N \times N$ coupling matrix

$$\hat{\Lambda} = \begin{cases} 1, & \text{if } m = n; \\ \lambda_{|m-n|} = \lambda_0 |m-n|^{-3}, & \text{if } m \neq n, \end{cases} \quad (11.4)$$

with λ_0 being the coupling coefficient between nearest neighboring SQUIDs.

In normalized form Eq. (11.3) reads

$$\ddot{\phi}_n + \gamma\dot{\phi}_n + \beta \sin(2\pi\phi_n) = \sum_{m=1}^{N}\left(\hat{\Lambda}^{-1}\right)_{nm}(\phi_{ext} - \phi_m), \quad (11.5)$$

where the frequency and time are normalized to $\omega_0 = 1/\sqrt{LC}$ and its inverse ω_0^{-1}, respectively, while fluxes and currents are normalized to Φ_0 and I_c, respectively, $\beta = I_c L/\Phi_0 = \beta_L/2\pi$ is the SQUID parameter, the overdots denote a derivation with respect to the normalized temporal variable, τ, and $\phi_{ext} = \phi_{ac}\cos(\Omega\tau)$, with $\Omega = \omega/\omega_0$ being the normalized driving frequency. The value of β_L determines whether a SQUID is hysteretic or nonhysteretic ($\beta_L > 1$ and $\beta_L < 1$, respectively).

Many times when the units are not very close to each other, it is reasonable to assume that the only nearest neighbor term in the longer range interaction plays a significant role. Even though originally it was thought that chimeras exist only for long-range interactions, we report below that also when the nonlinear features of the SQUIDs are very strong we may have chimeras. In order to find the chimera states, we need to integrate numerically Eqs. (11.5) in time using a fourth order Runge-Kutta algorithm with fixed time stepping, typically with $\Delta t = 0.02$ and initial conditions $\phi_n(\tau = 0)$ randomly chosen from a flat, zero mean distribution in $[-\phi_R/2, +\phi_R/2]$ and $\dot{\phi}_n(\tau = 0) = 0$ for all n. The boundary conditions $\phi_0 = \phi_{N+1} = 0$ and $\dot{\phi}_0 = \dot{\phi}_{N+1} = 0$ are used to account for the termination of the structure. For most initial flux configurations the system reaches spontaneously dynamical states where synchronous (coherent) and asynchronous (incoherent) clusters of SQUIDs coexist. A typical spatiotemporal flux pattern obtained after 10^7 time units (t.u.) of integration is shown in Fig. 11.2, where the evolution of the

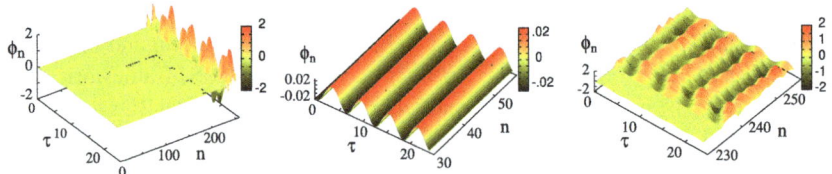

Fig. 11.2 Spatiotemporal evolution of the fluxes ϕ_n threading the SQUID rings during four driving periods $T = 5.9$ for $N = 256$, $\gamma = 0.0022$, $\lambda_0 = -0.05$, $\beta_L = 0.7$ ($\beta = 0.1114$), $\phi_{ac} = 0.015$, and $\phi_R = 0.85$. Left: for the whole SQUID metamaterial, middle: for part of the metamaterial that belongs to the coherent cluster, and right: for part of the metamaterial that includes the incoherent cluster

ϕ_ns is monitored for four driving periods $T = 2\pi/\Omega$. In the left panel, two different domains of the array can be distinguished, where the fluxes are oscillating either with low or with high amplitude. The enlargement of two particular subdomains in the middle and right panels reveals the main feature of a chimera state; besides the difference in the oscillation amplitudes, there are different dynamic behaviors: The low-amplitude oscillations are completely synchronized, while the high-amplitude ones are desynchronized, both in phase and in amplitude.

The degree of synchronization for a given cluster or for the whole SQUID metamaterial having M elements is quantified by introducing a Kuramoto-type parameter

$$\Psi = \frac{1}{M} \sum_{m=1}^{M} e^{i(2\pi \phi_m)}, \tag{11.6}$$

which provides a global measure of synchronization. In accordance with Eq. (11.6), the absolute value of the synchronization parameter, $|\Psi|$, lies in the interval $[0, 1]$, where the boundaries 0 and 1 correspond to complete desynchronization and synchronization, respectively. Perfect synchronization (i.e., $|\Psi| = 1$), however, cannot be achieved for finite systems like those considered here. In addition to the global synchronization parameters Ψ, we may also introduce local ones that give more detailed local information on the behavior of neighborhoods of oscillators. For a different type of chimera states that emerge from a variety of initial conditions and are shown in the left panels of Fig. 11.3, the corresponding *local synchronization parameter* $|Z_n|$ is shown in the right panels of the same figure. The real-valued local synchronization parameter $|Z_n|$ is a measure that can be calculated for a group of 2δ coupled oscillators (a subsystem of a larger system of N coupled oscillators) at every instant of time. Its value indicates the instantaneous degree of spatial coherence, i.e., the instantaneous degree of synchronization, of that group. It is defined as the magnitude of the complex parameter

$$Z_n = \frac{1}{2\delta} \sum_{|m-n| \leq \delta} e^{2\pi i \phi_m}, \quad n = \delta + 1, \ldots, N - \delta. \tag{11.7}$$

11.1 Introduction

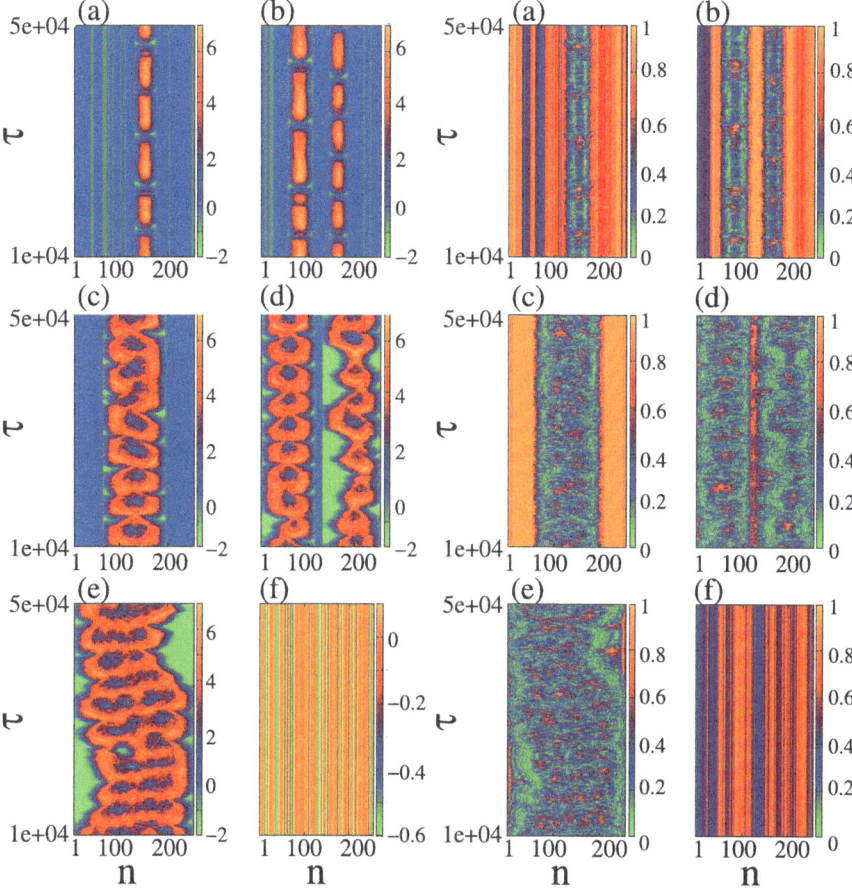

Fig. 11.3 Left: Space-time plots for the flux density ϕ_n of the SQUID metamaterial for different initial conditions. Panels (**a**) and (**c**) show chimera states with one desynchronized region, panels (**b**) and (**d**) show chimera states with two desynchronized regions, while panel (**e**) shows a state with a drifting desynchronized domain, and panel (**f**) shows a pattern with solitary states. Right: The corresponding space-time plots for the local order parameter $|Z_n|$ for the states are shown in the left panels. Parameter values are $T = 5.9$, $N = 256$, $\gamma = 0.0021$, $\lambda_0 = -0.05$, $\beta_L \simeq 0.7$, $\phi_{ac} = 0.015$, and $\phi_{dc} = 0.0$

A value of the local order parameter $|Z_n| = 1$ ($|Z_n| < 1$) indicates that the n-th oscillator belongs to a synchronized (desynchronized) cluster of the system of N oscillators.

The local order parameter is employed to quantify in each lattice neighborhood the degree of synchronization of the collective states obtained for the SQUID metamaterial. In the spatiotemporal flux patterns shown in the left panels of Fig. 11.3, the values of the ϕ_ns are obtained at time instants that are multiples of the driving period $T = 2\pi/\Omega$ of the ac flux field. In particular, Fig. 11.3a and c corresponds to typical chimera patterns which exhibit a cluster of desynchronized SQUIDs. In Fig. 11.3a this cluster is small, and it is located around $n = 150$, while in Fig. 11.3c it is much larger, spanning the region from $n \simeq 70$ to $n \simeq 190$. The SQUIDs that do not belong to these clusters are not all synchronized to each other. Instead, small subclusters of SQUIDs are apparent as stripes with uniform colorization. The SQUIDs that belong to such a stripe are synchronized; however, the stripes are not synchronized to each other. Furthermore, the flux oscillations in the SQUIDs that belong to the desynchronized clusters are much stronger than those in the other SQUIDs. In Fig. 11.3b and d, chimera states exhibiting two desynchronized clusters are shown. In Fig. 11.3b these two clusters are small, and they are located around $n \simeq 80$ and $n \simeq 160$. In Fig. 11.3d, on the other hand, the two desynchronized clusters are so large that do not leave any space for any synchronized cluster to exist. A drifting pattern can be observed in Fig. 11.3e, where the largest part of the SQUID metamaterial forms a desynchronized cluster, the size and position of which vary in time. Finally, Fig. 11.3f demonstrates a pattern of low-amplitude flux oscillations with multiple so-called solitary states, where many SQUIDs have escaped from the main synchronized cluster and perform oscillations of higher amplitudes (depicted by the light green stripes in the otherwise orange background). The degree of synchronization within the aforementioned states is visualized through the corresponding space-time plots of the local synchronization parameter, Eq. (11.7) shown in the right panels of Fig. 11.3. Red-orange colors denote the synchronized or coherent regions and blue-green colors the desynchronized or incoherent ones. These plots reveal the complexity of the synchronization levels in the SQUID metamaterial and show how an intricate balance between order and disorder may emerge in seemingly simple classical nonlinear systems.

11.1.1.1 SQUID Metamaterials with Local Coupling

Chimera states have mostly been found for nonlocal coupling between the coupled oscillators. This fact has given rise to a general notion that nonlocal coupling is an essential ingredient for their existence. This is also shown in the SQUID metamaterial where long-range coupling appeared to be essential for the creation of chimeras. However, if the metamaterial is operated in a strongly nonlinear regime, short-range interaction induced chimeras are also possible [10]. The emergence of multiclustered robust chimera states in locally coupled SQUID metamaterials can be demonstrated in an experimentally relevant parameter region. In Fig. 11.4 we see time-snapshots of the fluxes ϕ_n for different initial conditions and for two values of the loss coefficient γ which differ by an order of magnitude. The left panel is for $\gamma = 0.024$. The initial "sine wave" flux distribution for each simulation is shown by the gray solid line. The SQUIDs that are prepared at lower values form the coherent clusters of the chimera state, while those that are initially set at higher flux values oscillate incoherently. Moreover, as the "wavelength" of the initial flux distribution increases, so does the chimera state multiplicity, i.e., the number of coherent and/or incoherent regions. A similar behavior is observed for lower values of the loss

11.1 Introduction

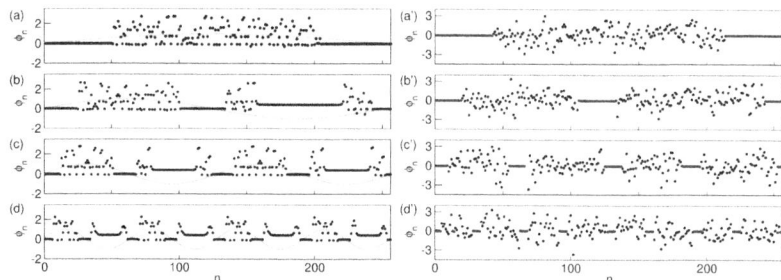

Fig. 11.4 Magnetic flux density ϕ_n at time $\tau = 5000$ time units for two different values of the loss coefficient: $\gamma = 0.024$ in (**a**)–(**d**) and $\gamma = 0.0024$ in (**a'**)–(**d'**). Gray solid lines mark the initial magnetic flux distribution used in the simulations. Blue solid lines in the right panel emphasize the coherent clusters of the chimera states. The other parameters are $T = 6.24$ ($\Omega \simeq 1.007$), $\beta_L = 0.86$, and $\phi_{ac} = 0.06$

coefficient $\gamma = 0.0024$ as shown in the right panel of Fig. 11.4. The incoherent clusters are better illustrated since they are approximately of equal size and do not contain oscillators that "escape" from the incoherent cluster abiding around low magnetic flux values, something which is visible in the left panel. Furthermore, the coherent clusters (emphasized by the blue solid lines) are fixed around $\phi = 0$, unlike in the left panel where additional clusters located at slightly higher values also form. Here we must recall that the snaking resonance[1] curve of a single SQUID increases significantly its winding with decreasing values of γ (right panel), creating thus new branches of stable (and equally unstable) periodic (period-1) solutions.

These branches are larger in number and smaller in size compared to those of higher γ values (left panel). The lower amplitude branches that are the biggest ones attract the SQUIDs that eventually form the coherent clusters. The other SQUIDs have multiple higher flux amplitude states to choose from and, therefore, create a more chaotic incoherent cluster than in the case of higher γ values. The observed chimera states can be quantified again through the local synchronization parameter $|Z_n|$ which is a measure for local synchronization [11].

11.1.2 Learning and Predicting Chimeras

Chimera states are surprising spatiotemporal patterns in which regions of coherence and incoherence coexist. While we focused so far on SQUID chimeras, there is a large number of biological, physical, engineering, etc. systems where chimeras appear, not only theoretically but also experimentally. The emergence of long-lived

---

[1] A nonlinear resonant behavior of a single SQUID at strong nonlinearity showing a snake-like figure [11].

chimera states in SQUID metamaterials can be achieved by proper initialization of the fluxes threading the SQUID rings, which is experimentally feasible. The numerical simulations rely on a realistic model, which is capable of reproducing experimental findings such as the tunability patterns of SQUID metamaterials which are obtained by varying an applied dc flux. These simulations indicate that in many cases the coherent and incoherent clusters that make a chimera state maintain their individual sizes for very long times ($> 10^8$ time units). During the integration, induced instabilities may lead to sudden expansion of the incoherent cluster(s) or to the desynchronization of of narrow coherent clusters. Moreover, other types of initial conditions may produce varying chimeras, where an incoherent cluster changes its position and size all the time.

A challenge for nonlinear dynamics is to predict future evolution of complex systems. While predicting the future evolution of periodic systems is relatively simple, this is definitely not true in quasiperiodic, chaotic, or spatiotemporally complex systems. More specifically, if we are able to device a method to predict reasonably well chimera states, we may use these techniques in other complex problems as weather patterns, climate, earthquakes, socially interacting networks, etc. We may thus see the problem of predictability of chimeras through machine learning as a test bed for applying AI motivated methods in many other useful and complex problems. Not only we can use chimeras to see if AI works but also we can compare different machine learning methods and see their relative strengths and weaknesses when applied in complex dynamics. To that effect we apply three different methods of machine learning and compare their predictive power when applied to spatiotemporally complex networks of oscillators. The methods we apply are the basic Feed-Forward Neural Network system FNN and the more sophisticated ones of Long Short-Term Memory (LSTM) and the reservoir computing (RC) recurrent neural network architectures. The LSTM networks have proven to be successful in predicting sequence-involving tasks such as speech recognition, machine translation, and human dynamics among others. While they have the ability to learn and reproduce long, isolated sequences, they are not necessarily able to provide robust predictions in complex physical systems that involve chaotic behavior or to capture dependencies between multiple correlated sequences. The RC network that comprises a linear input layer, a recurrent nonlinear reservoir network, and a linear output layer has ability to capture a large spectrum of frequencies embedded in complex systems. This approach has been applied successfully to inference problems in chaotic systems however with the addition of "observers" that provide continual "ground truth" for part of the system [12]. This method deduces the state of the chaotic dynamical system as a function of time from the limited number of the concurrent system state measurements. At each time step, the RC estimates the desired unmeasured variables from the measured variables and predicts the evolution of the physical system. Lu et al. have recently demonstrated the effectiveness of the method by applying it to the Rössler system, the Lorenz system, and the spatiotemporally chaotic Kuramoto-Sivashinsky equation, expressing the view that the method addresses the inference of unmeasured state variables rather than their prediction [12].

11.1.3 Observer-Based Neural Network Learning and Prediction

The LSTM networks have proven successful in learning and generalizing sequential tasks from isolated sequences, such as handwriting and speech. Inspired by this success, we first consider a model with a single LSTM network assigned to each system node, which is independent of all other LSTMs. The prediction error for this approach turned out to be very large. This is not surprising since chimera states are collective phenomena and the simplistic use of one LSTM network per node, independent of all others, does not capture well the interaction between the nodes; the sequences of different nodes are not isolated but correlated. In order to address the independent LSTMs' limited ability to capture dependencies between multiple correlated sequences, we add "observers" to the method and introduce the "Observer LSTM (OLSTM)" approach. This idea is an extension of the notion of "reservoir observers" to the LSTM networks. In the OLSTM method, we not only assign one LSTM to each (non-observer) system node but also assign "observer" status to certain system nodes ("LSTM observers," taken at equidistant positions for simplicity) which provide continual "ground-truth" measurements as input to the prediction method. Their presence even in small numbers, i.e., typically of order less than 10% of the total number of nodes, greatly improves the long-term data-driven forecasting capability of the LSTM networks and provides the framework for a consistent comparison between the RC and LSTM methods.

Time-series data are used to train each network, while no knowledge of the underlying system equations is required. Each individual LSTM network is trained by taking as input a number of N_p past values for the node at hand, plus the ground-truth values provided by all observers. Thus, the OLSTM method produces a generalized sequence, which combines the N_p past values and the time-varying systemic interaction. Using the trained networks, long-term predictions are made by iteratively predicting one step forward for each node, using as input the node's previous values and the values provided by the relatively very small number of observers. We evaluate the long-term forecasting capability of the OLSTM network as a function of the number of observers and as a function of training set size, and we compare the OLSTM performance with that of "reservoir observers" trained by RC, which utilizes a single ("global") network for the entire system [13]. Both OLSTM and RC methods are compared to standard feed-forward neural network method, with the same number of observers (OFNN). We compare quantitatively the networks' performance by calculating the normalized root mean square error (RMSE) at each time step, for all system nodes, over the predicted time steps, without counting observer nodes and training time steps, since they do not contribute to the prediction error.

We apply RC, OLSTM, and OFNN methods for long-term prediction of the dynamics of SQUID chimeras. Prediction snapshots for the single-headed chimera are presented in the right panels of Fig. 11.5. In predicting the evolution of this chimera, 17 "observers" were used in the positions marked by the tips of the arrows. The predicted time series for the flux of SQUID 163 is being shown (Fig. 11.5, right

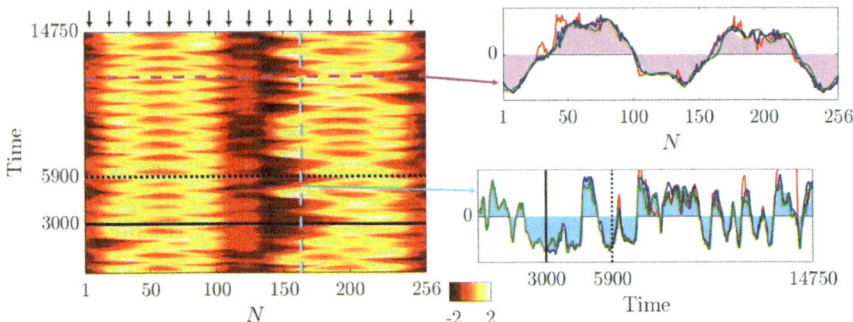

Fig. 11.5 Spatiotemporal plots of a double-headed chimera and predicted time series and fluxes generated in a one dimensional SQUID array depicting the evolution of the fluxes (values are color-coded) in a large array of 256 coupled SQUIDs by dipole-dipole moments, which has been studied numerically for a nonlocal coupling scheme. In predicting the spatiotemporal evolution of this chimera, 17 "observers" have been placed in the positions marked by the tips of the arrows. The thick horizontal black line marks the end of the RC training time, while the dotted horizontal black line marks the end of the OLSTM (and OFNN) training time. Right, top figure: predicted fluxes for all SQUIDs (entire metamaterial) at time step $t_n = 12,000$, red line depicts prediction by OLSTM, green line depicts prediction by RC, and blue line depicts prediction by OFNN. Right, bottom figure: Predicted time series for the flux of SQUID 163: Symbols are the same as in the right top figure, but with pink depicting the actual (ground-truth) data

bottom panel) and the predicted fluxes for all SQUIDs, i.e., the entire metamaterial, at time step $t = 12,000$ (Fig. 11.5, right top panel). The specific SQUID node 163 has been chosen in order to depict the long-term forecasting capability of the ML methods in challenging regimes that include both coherent and incoherent behavior. These are very long-time predictions, not just short-term prediction of just a few time steps beyond the training time; specifically, these predictions comprise more than twice (for OLSTMs and OFNNs) and three times (for RC) the training time. Nevertheless, the ML methods produce non-divergent predictions.

In the implementation of OLSTMs we found that a large value of the number of past steps N_p is not necessary, and in fact even with $N_p = 1$ the results are quite satisfactory; this implies that the presence of observers more than compensates the need for a short memory to guide the predictions. These results demonstrate that the "neighborhood" of the node is important; it changes dynamically, and the assistance of observers in predicting the future values is more appropriate to process the temporal variation of the "neighborhood" state. The same considerations apply to both SQUID-array and laser-array chimeras detailed in the next subsection.

11.2 Turbulent Chimeras

Chimeras can be stationary or turbulent [14]. Turbulent chimeras have been observed experimentally and have been classified in numerical studies of large arrays of SQUIDs and in arrays of lasers with various types of interactions. Their actual trajectories are highly nonlinear and comprise an immense challenge to predicting their occurrence. In the following section, we present the long-term prediction results of the machine learning methods applied on turbulent chimeras in simulated semiconductor laser arrays.

11.2.1 Coupled Semiconductor Lasers

In coupled semiconductor lasers we have a nonlinear lattice that can also be thought as a "metamaterial." The system generates numerous nonlinear modes including a certain type of chimeras that have additional complexity compared to the stationary SQUID chimeras. These are turbulent multiclustered chimeras that emerge in a large array of coupled semiconductor lasers with nonlocal coupling [10]. This array is a ring of $M = 200$ semiconductors lasers of class B. Each node j is symmetrically coupled with the same strength to its R neighbors on either side with nonlocal coupling.

The evolution of the slowly varying complex amplitudes $\mathcal{E}_j = E_j \exp\{(i\phi_j)\}$ (where E_j is the amplitude and ϕ_j the phase of the electric field) and the corresponding population inversions N_j are given by

$$\frac{d\mathcal{E}_j}{dt} = (1+ia)\mathcal{E}_j N_j + \frac{ke^{-i2C_p}}{2R} \sum_{l=j-R}^{j+R} \mathcal{E}_l \quad (11.8a)$$

$$\frac{dN_j}{dt} = \frac{1}{T}\left(p - N_j - (1+2N_j)|\mathcal{E}_j|^2\right), \quad j = 1,\ldots,M, \quad (11.8b)$$

where all indexes have to be taken modulo M. The parameter T is the ratio of the lifetime of the electrons in the excited level and that of the photons in the laser cavity. Lasers are pumped electrically with the excess pump rate $p = 0.23$. The linewidth enhancement factor a models the relation between the amplitude and the phase of the electrical field. We consider $a = 2.5$, which is a typical value for semiconductor lasers. The coupling strength k and the phase C_p are the control parameters that are used to tune the collective dynamics of the system. This complex coupling coefficient models the important effect of a phase shift introduced as the electric field of one laser couples into another. Equation (11.8) is a reduced form of the Lang-Kobayashi model in the limit where the delay of the external cavity tends to zero [10]. By shifting the coupling phase to $(C_p + \pi)$, we can obtain the model that describes the interaction of each field of semiconductor lasers in an array of waveguides where all indices have to be taken modulo M and T is the ratio of the lifetime of the electrons in the excited level and that of the photons in the laser cavity. Lasers are pumped electrically with the excess pump rate $p = 0.23$. The linewidth enhancement factor models the relation between the amplitude and the phase of the electrical field; a value of $a = 2.5$ was used typical for semiconductor lasers. The coupling strength k and the phase C_p are the control parameters that are used to tune the collective dynamics of the system.

11.2.1.1 Measures for Phase and Amplitude Synchronization

By using polar coordinates the characterization of the phase synchronization of our system can be done through the Kuramoto local order parameter $|Z_j|$ defined previously. We use a spatial average with a window size of $\zeta = 3$ elements. A Z_j value close to unity indicates that the j-th laser belongs to the coherent regime, whereas Z_j is closer to 0 in the incoherent part. This quantity can measure only the phase coherence and gives no information about the amplitude synchronization of the electric field. For the latter, we will use the classification scheme presented in [14] for spatial coherence. In particular, we may calculate the so-called *local curvature* at each time instance, by applying the absolute value of the discrete Laplacian $|DE|$ on the spatial data of the amplitude of the electric field:

$$|DE|_j(t) = \left| E_{j+1}(t) - 2E_j(t) + E_{j-1}(t) \right|, \quad j = 1, \ldots, M. \quad (11.9)$$

In the synchronization regime the local curvature is close to zero, while in the asynchronous regime it is finite and fluctuating. Therefore, if g is the normalized probability density function of $|DE|$, $g(|DE| = 0)$ measures the relative size of spatially coherent regions in time. For a fully synchronized system $g(|DE| = 0) = 1$, while for a totally incoherent system it holds that $g(|DE| = 0) = 0$. A value between 0 and 1 of $g(|DE| = 0)$ indicates the coexistence of synchronous and asynchronous lasers.

Note that the quantity g is time dependent. Complementary to the local curvature, we also calculate the spatial extent occupied by the coherent lasers which is given by the following integral:

$$g_0(t) = \int_0^\delta g(t, |DE|) d|DE|, \quad (11.10)$$

where $\delta = 0.001$ is a threshold value distinguishing between coherence and incoherence, which is related to the system-dependent, maximum curvature.

In the synchronization regime the local curvature is close to zero, while in the asynchronous regime it is finite and fluctuating. Since our objective is to obtain "wild" turbulent chimeras in order to test the long-term predictions of the ML methods, we have chosen a turbulent chimera regime generated with the choice of parameter values: $R = 64$, $C_p = 0.4\pi$, $k = 0.225$, $T = 392$, $p = 0.23$, and $a = 2.5$ [10]. Prediction snapshots are presented in Fig. 11.6 on the right-hand side as additional plots comparing with the actual (ground-truth) spatiotemporal

11.2 Turbulent Chimeras

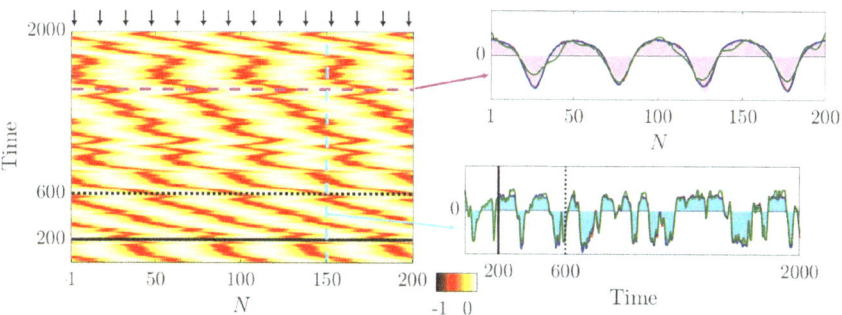

Fig. 11.6 Spatiotemporal plot of a turbulent chimera state in a one dimensional semiconductor Class B laser array depicting the evolution of the local curvature for each laser (values are color-coded), studied numerically for a nonlocal coupling scheme. Seventeen "observers" have been placed in the positions marked by the tips of the arrows. The thick horizontal black line represents the size of the training dataset used for training the reservoir observers with RC, while the dotted horizontal black line represents the size of the training dataset for training the OLSTM (and OFNN) observers. Right, top trace: snapshot of the spatial profile of the predicted local curvatures for the entire array at time step $t_n = 1500$. Shaded blue color depicts the actual (ground-truth) data, red line depicts prediction by OLSTM, green line depicts prediction by RC, and blue line depicts prediction by OFNN. Right Bottom trace: the predicted time series for laser 145, the color code is as in the top right figure, but with pink color depicting the actual (ground-truth) data

evolution of the local curvature for the flux of laser 145 and for all lasers, for similar choices as in Fig. 11.5, that is, with 17 "observers" and at time step $tn = 1,500$. These are extremely long-time predictions, with time horizons almost five times the training time (for OLSTMs and OFNNs) and ten times (for RC) the training time. In this case, as in that of the SQUID system, all ML methods under study produce non-divergent predictions.

In Fig. 11.7 we present the calculated values of the normalized root mean squared error for long-term prediction (top inset) of the RC, OLSTM, and the OFNN methods and as a function of the number of observers and the size of the training dataset for a randomly selected time step for the chimera depicted in Fig. 11.6; similar results are obtained for all systems. The RMSE values remain, on average, close to 101, for time horizons up to 2,000 time steps in the future. This figure also presents graphs of how the RMSE varies as a function of the number of observers (given as percentage of the number of the system nodes) and as a function of the size of the training datasets (given as percentage of the entire ground-truth time series). In all cases, low RMSE values are attained after a minimum number of observers have been included in the system, comprising about 5% of the total number of oscillators, and the size of the training data reaches at least 15% of entire dataset. These results demonstrate that even very small numbers of observers are very important for achieving non-diverging long- term predictions. Furthermore, they show that RC achieves low levels of RMSE with smaller size of training datasets, whereas OLSTM achieves low levels of RMSE with a smaller number of observers.

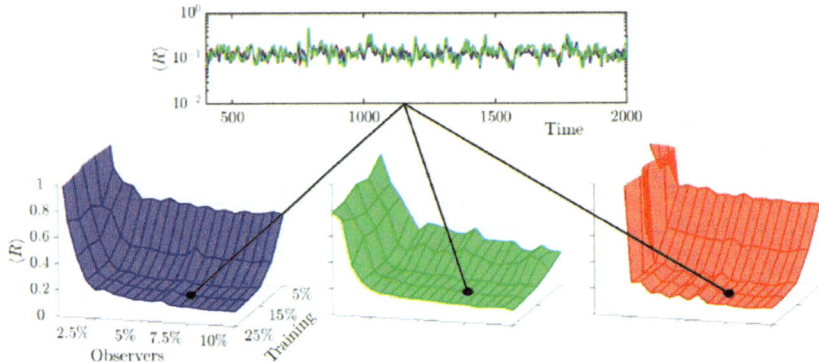

Fig. 11.7 Normalized RMSE ($\langle R \rangle$) of RC, OLSTM, and the OFNN methods calculated for each time step over all system nodes at each predicted time step (top panel) and over the predicted time steps and all system nodes (bottom three dimensional plots) as a function of the number of observers (given as percentage of the entire number of system nodes) and the size of the training dataset (given as percentage of the entire ground-truth time series) for the chimera depicted in Fig. 11.5. The black dots in each of the three dimensional plots represent the normalized RMSE for each ML method, respectively, at a randomly picked time step, calculated overall system nodes and previous time steps, but not counting observer nodes and training time steps. Red: OLSTM, green: RC, and blue: OFNN. Axes, tick marks, tick labels, and grid lines are the same in all three dimensional plots

11.2.1.2 Parameter Regimes

For the machine learning methods we used specific tools with certain parameter values. The RC network architecture used in predicting turbulent chimeras comprises 1,000 reservoir nodes, with spectral radius $\rho = 1.0$, average degree $D = 80$, scale of inputs weights $\sigma = 1.5$, bias constant $\xi = 0.0$, leakage rate $\alpha = 0.9$, ridge regression parameter $\beta = 0.5$, and time interval $dt = 0.01$. The RC network applies to the system as a whole (single network architecture).

The OLSTM network architecture used constitutes 400 LSTM cells with *ReLU* activation functions and one hidden layer. A single LSTM network is applied to each non-observer system node that is either a SQUID or a laser oscillator. Similarly, a single fully connected, dense feed-forward network (OFNN), with *ReLU* activation functions, is applied to each non-observer system node, SQUID or laser oscillator. Optimization for LSTMs during training is performed using the Adam stochastic optimization method with a learning rate of 0.001 as implemented in Keras.

In case of the SQUIDs system (Fig. 11.1), the OLSTM and OFNN models were trained with 5900 time steps (40% of the total 14,750 time steps) and RC with 3000 time steps (about 20% of the total number of time steps). In case of the laser chimeras system of Fig. 11.6, the OLSTM and OFNN models were trained with 400 time steps (20% of the total number of time steps and RC model with 200 time steps (10% of the total number of time steps). Finally, in case of branching (Fig. 11.4), OLSTM and OFNN were trained with 352 time steps (40% of the total number of time steps) and RC with 300 time steps (about 34%). The time-series data have been divided into two separate sets, the training dataset and the validation dataset. The data is stacked in batches (of size 50 data points) in order to form the training (and validation) input and output of the networks. These training batches are used to optimize the parameters of the networks (weights and biases). The training proceeds by optimizing the network weights iteratively for each batch. The training loss function is a weighted version of the root mean square error. When the network parameters have been optimized once for all training data batches, one epoch is completed. After every epoch the RMSE in the validation dataset is computed. Training is stopped after 200 epochs, and the OLSTM (OFNN) network with the smallest validation error is selected in order to avoid overfitting.

Each trained OLSTM network is then used to forecast the node's state in the next time step in an iterative fashion (getting also input from the observers). All the hyperparameters in OLSTM and OFNN models including the number of training epochs, the number of training batches, the number of neurons, and the number of hidden layers are optimized so that they are leading to the smallest RMSE. Similarly, for RC we optimized the parameters with criterion the smallest value for RMSE. As a comparison measure for the networks' performance, we use the normalized root mean square error calculated at each time step, for all system nodes and over the predicted time steps.

11.2.2 Random and Moving Observers

We close this chapter on chimera predictions by focusing on the robustness of the long-term forecasting capability of observer based methods, viz. RC, OLSTM, and OFNN. Since all three of them appear to give reasonably close predictions in a compatible regime, we will simply use OFNNs that train much faster and apply them to the spatiotemporal evolution in the optical system that involves turbulent chimeras. The OFNNs method we employ assigns one network to each one of the system's nodes except for the "observer" nodes which provide continual "ground-truth" measurements as input. In order to investigate the robustness of the prediction capabilities of the OFFNs method in respect of different observers placements, we use the following observers' schemes: (a) uniformly distributed equidistant observers at all time steps, (b) randomly assigned observers at each time step, (c) uniformly distributed equidistant observers moving over time in phase (performing oscillatory periodic motion around their positions, with amplitude equal to five-node distance), in both training and testing, (d) randomly assigned observers, at fixed positions for all time steps, (e) randomly assigned observers at fixed positions for all time steps, moving in phase around their fixed positions in an oscillatory periodic motion with amplitude equal to five-node distances, (f) uniformly distributed equidistant observers, stationary during training, but moving in phase during testing—the oscillatory periodic motion around their positions, with amplitude equal to five-node distance, and (g) uniformly distributed equidistant

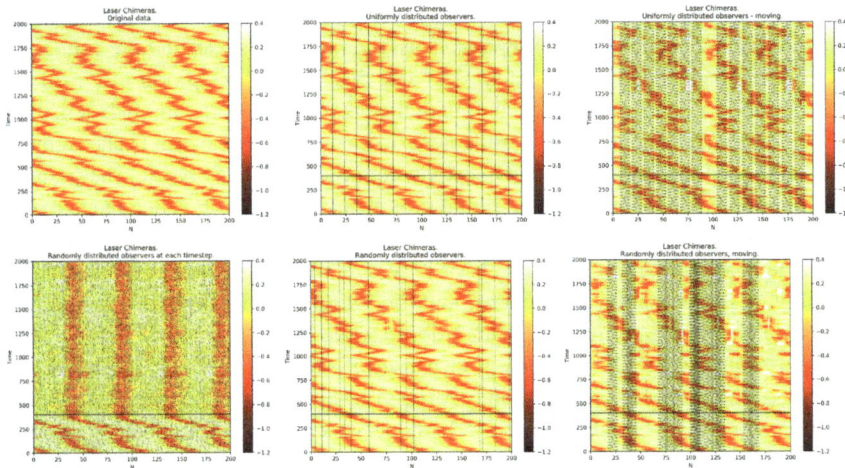

Fig. 11.8 Spatiotemporal plots of the turbulent chimera state generated in [13] Figure 4V. (**a**) The evolution of the chimera state as studied numerically in [13] for a nonlocal coupling scheme, (**b**) results obtained by incorporating uniformly distributed (equidistant) observers at all time steps, (**c**) results obtained by incorporating randomly assigned observers at each time step, (**d**) results obtained by incorporating uniformly distributed (equidistant) observers moving over time in phase (performing oscillatory periodic motion around their positions, with amplitude equal to five-node distance), in both training and testing, (**e**) results obtained by incorporating randomly assigned observers (at fixed positions for all time steps), (**f**) results obtained by incorporating randomly assigned observers at fixed positions for all time steps, moving (in phase) around their fixed positions in an oscillatory periodic motion (amplitude equal to five-node distance), (**g**) results obtained by incorporating uniformly distributed (equidistant) observers, stationary during training, but moving (in phase) during testing (oscillatory periodic motion around their positions, with amplitude equal to five-node distance), and (**h**) results obtained by incorporating uniformly distributed (equidistant) observers during training, but moving (with random phase) during testing

observers during training, but moving with a random phase during testing [15]. Although there are many other observed distributions, the static and moving observers we chose to investigate practically exhaust the interesting modes of interaction of observes with the AI-ML system. From these modes we arrive at quantitative conclusions on observer-assisted AI predictability of spatiotemporal systems (Fig. 11.8).

11.3 Conclusion

The issue of predicting complex spatiotemporal behavior using machine learning approaches is one of central importance for fundamental science and practical applications. In this chapter, we addressed this issue by considering a prototypical phenomenon, viz. partially coherent chimera states as these are realized in coupled arrays of SQUIDs or lasers. We find that ML approaches like LSTM and RC recurrent neural networks can perform well in predicting complex dynamics in extended physical systems when they involve "observers" that monitor the system evolution throughout its time dimension. The presence of observers is an inherent requirement of the second approach (RC), but not of the first (LSTM). Accordingly, the "Observer LSTM" may be used to address the limitations of single, independent LSTM networks in capturing dependencies between multiple correlated sequences. We have also considered an observer-enhanced feed-forward network (OFNN) and tested the long- term prediction performance of the three approaches, OLSTM, OFNN, and reservoir observers trained by RC, on the challenging problems of stationary and turbulent chimeras. The outcome of the analysis quantifies how the prediction error, i.e., the root mean squared error, RMSE, of the predicted values, varies as a function of the number of observers and of the size of the training datasets.

We conclude that sensors in the form of observers are of paramount importance for a robust data-driven long-term forecasting in complex systems. Observer-enhanced ML methods, like OLSTM, acting as high-level "intelligent" interpolation schemes, are capable of successfully predicting the nonlinear spatiotemporal evolution of complex dynamical systems. Many issues remain to be evaluated and questions to be answered in establishing the robustness and efficiency of the observer-enhanced approaches. The robustness of the forecasting capabilities of the OFNN models versus the distribution of the observers, including equidistant and random, and the motion of them, including stationary and moving, was also investigated. Especially in the case of turbulent chimeras the predictability of the model does not show strong dependence on the spatial distribution of the observers; both equidistant and randomly distributed observers are able to predict the behavior of the system. When the observers are moving in phase, the prediction is less accurate but still satisfactory. In cases where the model is trained with stationary uniformly distributed observers, but in the prediction stage the observers are moving around their initial position (oscillatory motion), the OFNNs are still able to predict the evolution of the system. However, the model fails in the case of random observers that are moving randomly at each step. In the case of the modular biological spatiotemporal state, the predictability of the model does not show strong dependence in either the spatial distribution of the observers or their motion. As before, the model fails in the case of random observers that execute a random walk. It is possible that the method has broader applicability in dynamical system context when partial dynamical information about the system is available.

11.4 Summary

- Chimeras are spatiotemporal structures that are coherent and chaotic.
- Machine learning methods may use observers to add real time ground-truth information to predict more efficiently.

- Reservoir computing, feed-forward neural networks, and long short-term memory recurrent neural networks with the help of observers learn to predict well chimera states.

References

1. K.K. Likharev, *Dynamics of Josephson Junctions and Circuits* (Gordon and Breach, Philadelphia, 1986)
2. N. Lazarides, G.P. Tsironis, rf SQUID metamaterials. Appl. Phys. Lett. **90**, 163501 (2007)
3. S.M. Anlage, The physics and applications of superconducting metamaterials. J. Opt. **13**, 024001 (2011)
4. P. Jung, S. Butz, S.V. Shitov, A.V. Ustinov, Low-loss tunable metamaterials using superconducting circuits with Josephson junctions. Appl. Phys. Lett. **102**, 062601 (2013).
5. P. Jung, A.V. Ustinov, S.M. Anlage, Progress in superconducting metamaterials. Supercond. Sci. Technol. **27**, 073001 (2014)
6. N. Lazarides, G.P. Tsironis, Superconducting metamaterials. Phys. Rep. **752**, 1 (2018)
7. Y. Kuramoto, D. Battogtokh, Coexistence of coherence and incoherence in nonlocally coupled phase oscillators. Nonlinear Phenom. Complex Syst. **5**, 380 (2002)
8. D.M. Abrams, S.H. Strogatz, Chimera states for coupled oscillators. Phys. Rev. Lett. **93**, 174102 (2004)
9. N. Lazarides, G. Neofotistos, G.P. Tsironis, Chimeras in SQUID metamaterials. Phys. Rev. B **91**, 054303 (2015)
10. J. Shena, J. Hizanidis, P. Hoevel, G.P. Tsironis, Multiclustered chimeras in large semiconductor laser arrays with nonlocal interactions. Phys. Rev. E **96** 032215 (2017)
11. J. Hizanidis, N. Lazarides, G.P. Tsironis, Robust chimera states in SQUID metamaterials with local interactions. Phys. Rev. E **94**, 032219 (2016)
12. Z. Lu, J. Pathak, B. Hunt, M. Girvan, R. Brockett, E. Ott, Reservoir observers: model-free inference of unmeasured variables in chaotic systems. Chaos **27**, 041102 (2017)
13. G. Neofotistos, M. Mattheakis, G.D. Barmparis, J. Hizanidis, G.P. Tsironis, E. Kaxiras, Machine learning with observers predicts complex spatiotemporal behavior. Front. Phys. **7**, 24 (2019)
14. F.P. Kemeth, S.W. Haugland, L. Schmidt, I.G. Kevrekidis, K. Krischer, A classification scheme for chimera states. Chaos **26**, 094815 (2016)
15. G.D. Barmparis, G. Neofotistos, M. Mattheakis, J. Hizanidis, G.P. Tsironis, E. Kaxiras, Robust prediction of complex spatiotemporal states through machine learning with sparse sensing. Phys. Lett. A **384**, 126300 (2020)

Chapter 12
Branching

AI for Complex Singular Events

Abstract When waves propagate in media with fluctuating index of refraction, they may merge and form branches of high-amplitude intensities in certain random directions. The resulting wave branching is a singular phenomenon that induces some form of coherence through randomness. It may appear in many physical, chemical, and geological and biological systems and has some general features that can be addressed analytically. In quantum physics particles such as electrons have wave-like properties and thus may show similar branching features. Graphene and more generally Dirac solids constitute two dimensional materials where the electronic flow is ultra-relativistic. When a Dirac solid is deposited on a substrate surface with roughness, a local random potential develops through an inhomogeneous charge impurity distribution that affects the charge flow. The result is a chaotic pattern of current branches that develops through focusing and defocusing effects produced by the random surface potential. An additional bias voltage may be used to tune the branching pattern of the currents. Analytical and numerical techniques can be employed in order to investigate the onset and the statistical properties of carrier branches in Dirac solids. We apply machine learning and evaluate the possibility of learning and predicting this statistical phenomenon. We find that methodology similar to that used in chimeras is able to capture the essence of the phenomenon.

12.1 Introduction

Extreme wave focusing due to refractive index variation is a common occurrence in many physical systems [1–4]. In optical media, for example, the index of refraction changes in a statistical way due to small imperfections or distributions of defects in the medium through which the wave propagates. Random spatial variability of the index leads to local focusing and defocusing of the waves and the formation of caustics that appear as "branches" with large local wave intensity [5]. Quantum particles like electrons also exhibit wave properties, and as a result, electrons traveling in disordered media can form coalescing trajectories and exhibit phenomena similar to wave motion in optical random media [1] (Fig. 12.1). Branching may have implications for technological applications, for instance, in

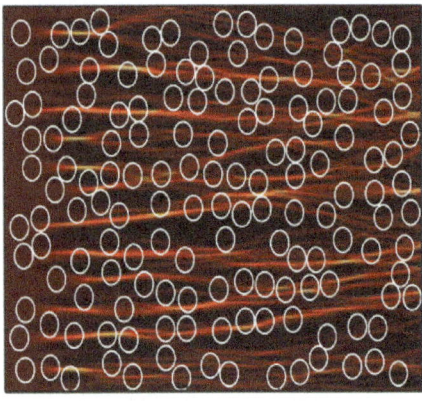

Fig. 12.1 Branching rays of light propagating in an artificial medium with a random distribution of (oval shape in the figure) Luneburg lenses. The random distribution of the effective index of refraction in the medium leads to ray coalescence and branching [3]

the ultra-relativistic electronic flow on a two dimensional (2D) random potential in graphene and other Dirac solids. In these cases, branching arises from the flow of electrons through a region of inhomogeneous distribution of charge impurities that are in the substrate and which create a random potential for the electrons. An additional bias voltage is introduced to induce the electronic propagation. The onset of the electronic branched flow is determined by the statistical properties of the random substrate potential, which is quantified through a scaling-type relationship to capture the emergence of branches. It is a formidable challenge to predict singular events like branching in wave propagation or electron flow because of the stochastic nature of the onset of such events [6]. An important question is whether machine learning methods can learn the interactions that take place among trajectories, thereby providing an accurate detection mechanism for the caustics that mark the onset of branching. We will present such a study using the methodology introduced in the previous Chap. 11 on chimeras and apply the RC, OLSTM, and OFNN methods on singular branched flows in graphene with random potentials. The results show that machine learning methods can capture the stochastic temporal dependencies of the time series in this prototypical complex dynamical system.

Wave focusing due to refractive index variation is a common occurrence in many physical systems. In sea waves, the effective refractive index variation arises from fluctuating depth while in optical media the index of refraction changes in a statistical way due to small imperfections or random distributions of defects. Random spatial variability of the index leads to local focusing and defocusing of the waves and the formation of *caustics* or wave *branches* with increased local wave intensity. Under general circumstances the branching flow develops a stochastic web with statistical patterns of persisting enhanced intensity wave motion. Since electrons have also wave properties due to their quantum nature, similar phenomena appear in the quantum realm. Injected electrons in disordered two dimensional electron gas form coalescing trajectories and manifest phenomena similar to wave motion in random media. One aspect of the electronic motion that has not yet been explored is the relativistic and in particular the ultra-relativistic one; the latter occurs in materials referred to as *Dirac solids* (DS), such as graphene. In the ultra-relativistic limit, the magnitude of the Dirac fermion velocity cannot be affected by external fields as it is already at its maximum value, leading to significant differences from the conventional nonrelativistic flow. This is directly reflected in the electronic branching properties and gives rise to discernible differences.

Pristine graphene is the prototypical two dimensional material, characterized by linear dispersion in the electronic band structure near the Fermi level

$$\epsilon_{\mathbf{k}} = v_F \hbar |\mathbf{k}|, \qquad (12.1)$$

where $\epsilon_{\mathbf{k}}$ is the single particle energy, v_F is the Fermi velocity, and $\mathbf{k}$ is the wave-vector. Electron flow in these band segments is ultra-relativistic with maximal propagation velocity v_F. This electron flow is modified by the presence of a bias potential applied along a specific direction. The relativistic electronic dispersion couples the motion of electrons along the direction of bias and the one perpendicular to it. Electron dynamics is also subject to the presence of substitutional or other type of weak disorder; the effects of such disorder are observed in graphene in the form of electronic puddles. The combined presence of disorder and bias alters the electronic flow, and thus the ultra-relativistic trajectories coalesce into branches of substantial local density.

12.2 Theoretical Formulation

We give a brief description of the basic formulation for obtaining analytically the onset of branches in graphene. This analysis helps to comprehend the basic physics of the onset of branching. The fundamental feature of a two dimensional Dirac solid such as graphene is the linear dispersion relation of the energy with wave-vector, Eq. (12.1). When the electronic density is low, we may use the independent electron model to describe the quasi-classical dynamics of charge carriers through the quasi-classical ultra-relativistic Hamiltonian

$$H = \pm v_F \sqrt{p_x^2 + p_y^2} + V(x, y), \qquad (12.2)$$

where $\mathbf{p} = (p_x, p_y)$ is the 2D momentum of the charge carriers. The Hamiltonian (12.2) is the classical limit of Dirac equation that describes dynamics of massless electron/hole quasiparticles in graphene and other Dirac solids. Branched flow is an effect of ray fields associated with waves, and thus, we may use the Hamiltonian (12.2) to give a ray description of the quantum flow of Dirac electrons. In the two dimensional ultra-relativistic dynamics the particles propagate in a medium with potential $V(x, y) = V_r(x, y) + V_d(x)$, where $V_r(x, y)$ is a random δ-correlated potential with energy scale much smaller than that of the electronic flow. The potential arises from random charged impurities in the substrate or other sources of disorder in the graphene sheet. A deterministic control potential $V_d(x) = -\alpha x$ is due to an externally applied voltage in the x-direction with α fixed by experimental conditions. Electrons are injected in the Dirac sheet with initial momentum p_0 along the x-direction; due to the ballistic electronic motion along the x-axis, we may ignore the effects of the random potential in this direction. For plane wave initial conditions for the electrons, i.e., $p_x(0) = p_0$ and $p_y(0) = 0$, we express the solutions of Hamilton's equations as $p_x(t) = p_0 + \alpha t$ and $p_y(t) = -\int \partial_y V_r(x, y) dt$, with $p_x \gg p_y$. We expand the Hamiltonian (12.2) up to second order in p_y/p_x and obtain

$$H = p_x + \frac{p_y^2}{2p_x} + V(x, y) \tag{12.3}$$

and set $v_F = p_0 = 1$ for simplicity. In order to study wave-like electronic flow through the Hamilton-Jacobi equation, it is best to introduce the classical action S with $p_x = \partial_x S$ and $p_y = \partial_y S$ and express the equation as

$$\partial S + \partial_x S + \frac{(\partial_y S)^2}{2\partial_x S} + V(x, y) = 0. \tag{12.4}$$

Employing previous non-relativistic approaches to the branching problem [1], we can derive an equation for the local curvature of the electronic flow determined through $u = \partial_{yy}$. To this effect one needs to apply the operator $\equiv (\partial_{xx} + \partial_{yy} + 2\partial_{xy})$ in Eq. (12.4) and obtain the following equation:

$$\partial_t u + \frac{u^2}{\partial_x S} + \frac{\partial_y S}{\partial_x S} \partial_y u + V(x, y) = 0. \tag{12.5}$$

Using the effective Hamiltonian (12.3) we calculate the equations of motion for the x, p_x conjugate variables keeping the lowest order terms in p_y/p_x, which leads to the simple relation $x(t) = t$. Thus, under the approximation of the dominance of the momentum in the forward x-direction, we find that space and time variables are identical. We use this fact to simplify the Hamilton-Jacobi ray dynamics and turn Eq. (12.5) into a quasi-2D version. This is done by replacing the space coordinate x with t and using $V = \partial_{yy} V_r(t, y)$. This simplification is valid only in the ultra-relativistic limit since nonrelativistic electrons accelerate in the presence of nonzero values for the parameter α.

We need to calculate the convectional derivative for u; for an arbitrary function $f(H)$, where H is the Hamiltonian (12.3), we have

$$\frac{df}{dt} = \left[\partial_t + \frac{\partial x}{\partial t}\partial_x + \frac{\partial y}{\partial t}\partial_y\right] f. \tag{12.6}$$

In the quasi-two dimensional approximation the potential depends only on time and transverse coordinate y, i.e., $V \equiv V(t, y)$; subsequently, the term that includes the operator ∂_x is zero. On the other hand, the term $\partial y/\partial t$ can be determined by Hamilton's equations of (12.3) as

$$\frac{\partial y}{\partial t} = \frac{dy}{dt} = \frac{\partial H}{\partial p_y} = \frac{p_y}{p_x}. \tag{12.7}$$

Using the definition of the classical action S, i.e., $p_x = \partial_x S$ and $p_y = \partial_y S$, and Eqs. (12.7) and (12.6), we obtain the convectional derivative formula:

$$\frac{df}{dt} = \left[\partial_t + \frac{\partial_y S}{\partial_x S}\partial_y\right] f. \tag{12.8}$$

Using the expression of Eq. (12.8) for the convectional derivative in conjunction with the approximate quasi-one dimensional Hamiltonian of Eq. (12.3), we obtain an ordinary nonlinear differential equation for the local wave curvature:

$$\frac{du}{dt} + \frac{u^2}{1+\alpha t} + \partial_{yy} V_r(t, y) = 0. \tag{12.9}$$

The dynamics of Eq. (12.9) determines the onset of the regime for caustics; this occurs at times, or equivalently locations along the x-axis, where the curvature u becomes singular. The first time when a singularity in u occurs determines the precise point for the onset of ray coalescence. Given that the term $\partial_{yy} V_r(t, y)$ is fluctuating, we may solve the first passage time problem for the curvature to reach $|u(t_c)| \to \infty$, where t_c is the time for the occurrence of the first caustic. The analytical solution to this problem determines a scaling law for the development of the first branching effect in the graphene sheet [4]. We proceed by employing numerics that produce a clear picture for the caustic generation in graphene.

12.3 Numerical Solution of the Hamilton-Jacobi Equation

In the numerical simulations we may directly use the characteristic equations for the full Hamiltonian (12.2) while constructing a random potential based on experimental observations. In particular, impurities in the substrate of graphene create a smooth landscape of charged puddles of radius $R \approx 4$ nm. In the model, each puddle size is drawn from a two dimensional Gaussian distribution with standard deviation R. The location of each puddle is randomly chosen through a uniform distribution. Extensive simulations have been performed for quasi-classical electron dynamics in a graphene sheet of size 400×400 nm, where several caustics are observed with periodic boundary conditions that ensure that all the rays reach a caustic [4]. The random potential consists of 2000 randomly distributed Gaussian defects with $R = 4$ nm. A collection of 1000 ultra-relativistic rays, initially distributed uniformly along y axis, are injected into the graphene sheet from the $x =$

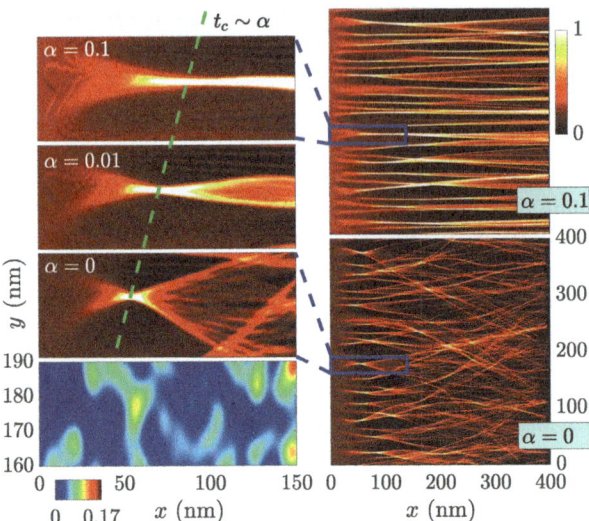

Fig. 12.2 Two dimensional numerical ray-simulations determine the onset of a caustic event in a disordered potential with $\sigma = 0.1$ and for a deterministic potential with $\alpha = [0, 0.05, 0.1]$. (Left) The lower panel shows the random potential. The remaining images represent the density of rays I. The green dashed line shows that the first caustic time t_c increases linearly with α. (Right) The ray density of branched flow in a graphene sheet for $\alpha = 0$ and $\alpha = 0.1$

0, with plane wave initial conditions, $p_x(0) = p_0 = 1$ and $p_y(0) = 0$. We select a single caustic event out of many in order to observe how the deterministic part of the potential affects the onset of this event, as in Fig. 12.2. The rays propagate in the disordered potential with $\sigma = 0.1$, and after time t_c a caustic event occurs. The ray-tracing simulations are performed for three different values of $\alpha = [0, 0.05, 0.1]$, to show that t_c increases linearly with α, a behavior expected also theoretically [4]. It is then confirmed numerically that the quasi-2D analytical prediction obtained from the analytical solution of the mean first passage time caustic problem in the presence of a small voltage in graphene *shifts* the location of the first caustic [4]. The location and the shape of caustics are modified by the external potential V_d; in particular, as α increases, the passage to branched flow is delayed and the caustics disperse slower, see Fig. 12.2.

12.4 Machine Learning

In the analysis that proceeded we presented the physical mechanism for branching phenomena that appear in two dimensional Dirac solids. The electronic flow is ultra-relativistic, while the presence of a weak disorder potential due to the substrate leads to branching. The latter persists even in the presence of an applied voltage

12.4 Machine Learning

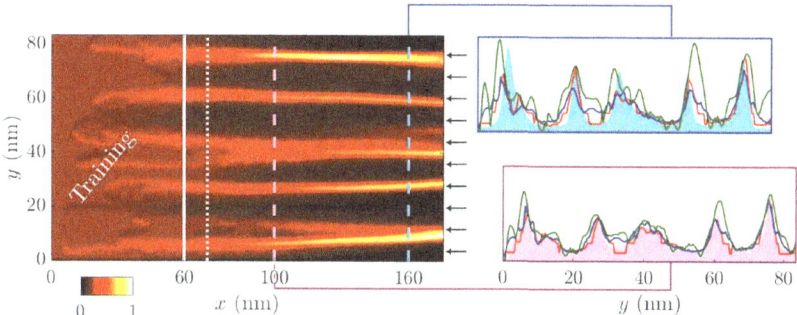

Fig. 12.3 *Left:* Branching electronic flow in graphene with random substrate potential with color-coded intensity. The graphene sheet has size 176 (vertical) × 84 nm (horizontal). The figures on the right depict the intensity of the flow as found in the simulation and predicted by the machine learning methods. We used ten observers located in the positions marked by the tips of the arrows, monitoring the entire "time" (vertical) axis. The thick horizontal white line marks the end of the RC training, while the dotted white line marks the end of the OLSTM (and OFNN) training. *Right:* The above figure outlined in blue shows the actual and predicted time series for the entire system at "time coordinate" point $x_n = 160$ nm, corresponding to 801 time steps, and with blue-shaded curve depicting the actual ground-truth data. The bottom figure outlined in pink shows the actual and predicted time series for the entire system at "time coordinate" point $xn = 100$ nm, corresponding to 501 time steps. The pink-shaded curve represents the actual ground-truth numerical data, the red line is the OLSTM prediction, green is the RC prediction, and blue is the OFNN prediction. *Line color coding:* Red: OLSTM, green: RC, and blue: OFNN

in the form of an external potential. This potential may reduce the dispersion of caustics and increase the time for occurrence of the first caustic event. While the onset of caustics and branching is essentially a stochastic phenomenon, we do apply machine learning methodology to it in order to see the extent of its power [6].

For the caustic event prediction in a 2D electron flow, we consider one of the spatial dimensions—we refer to it as the "longitudinal" x-direction—as the "time-coordinate" and, therefore, map the stationary phenomenon of caustic formation onto a 1D spatiotemporal dynamical problem. In the framework of this approach, the motion of electrons is modeled as individual rays whose density matrix is transformed onto a vector of time series with dimension N, the number of mesh points of the remaining spatial dimension, i.e., the "transverse," or y- direction. We use $N = 210$ that spans a total length of 84 nm. The goal is to predict the onset of branching in time and its location in the electronic flow. Figure 12.3 presents prediction snapshots of the onset of branching in electronic flows in graphene with random potentials, depicting the intensity of the flows, using ten different "observers" along the "time" axis of the flows at different positions on the transverse axis. These results demonstrate that the RC, OLSTM, and OFNN methods are able to predict well the branching in the electronic flows in this system, even at very long prediction times. In these methods, observers are very important to achieve long-term forecasting capability; they act as real time sensors that help predict the future values of the flows.

The networks we used are similar to the ones used for the prediction of chimera states in the previous Chap. 11. The OLSTM network consists of 400 LSTM cells with RELU activation functions (one hidden layer). A single LSTM network is applied to each (non-observer) system node. The single fully connected feed-forward network (OFNN), with RELU activation functions, is applied to each (non-observer) system node. Optimization for LSTMs during training is performed using the Adam stochastic optimization method with a learning rate of 0.001 as implemented in Keras. For branching, the reservoir computing network used comprises 3000 reservoir nodes, with spectral radius $\rho = 0.9$, average degree $D = 50$, scale of inputs weights $\sigma = 1$, bias constant $\xi = 0.4$, leakage rate $\alpha = 0.5$, ridge regression parameter $\beta = 0.05$, and $dt = 0.05$. The network applies to the system as a whole. The OLSTM and OFNN were trained with 352 time steps (40% of the total number of time steps) and RC with 300 time steps (about 34%).

12.5 Conclusion

The onset of wave caustics is essentially a random phenomenon due to fluctuations in the index of refraction in the medium. It applies equally well not only to classical waves and optics but also to quantum mechanical particles. The latter are also waves according to quantum mechanics and, as a result, can also form branches as they propagate in random media. The application of machine learning methods in this type of systems is in principle "risky," since how can we expect that a deterministic method based on neural networks may predict random events? On the other hand, it is possible that during the training phase the networks learn hidden correlations of a seemingly random medium and then can use this information for prediction. In order to test these ideas and the power of prediction of neural networks in random media, we used the phenomenon of branching in graphene and Dirac solids in general. Graphene is a unique two dimensional solid where the propagation of electrons is done with an ultra-relativistic dispersion, i.e., follow a Dirac cone. When placed on a substrate the electrons experience a fluctuation potential that leads to the generation of high-intensity branch lines in their propagation. This phenomenon can be analyzed rigorously mathematically and also can be shown numerically. It is a genuine emergent phenomenon that stems from the complexity of the material.

The use of three different methods of machine learning, viz. "simple" feed-forward neural networks, long short-term memory recurrent networks and reservoir computing showed that it is indeed possible to learn and thus predict the onset of branches on the graphene surface. All three methods worked relatively well when assisted by the presence of observers, i.e., sensors that give real time information in several system locations. As in the chimera prediction case, similarly in branching, there is a competition between the number of sensors used and the accuracy of the future evolution. When the number of observers is very small, the predictability horizon is very limited or in reality lost.

Can we then predict the future in complex spatiotemporal systems? The answer is conditional; we may do this efficiently if we additionally use partial but real time ground truth information. The network cannot predict without some "help" from the real system since locally the system may have diverging properties. How many observers do we need to predict well? This depends on the system and the horizon we want to predict. The observers turn the prediction process to a form of interpolation from a trained network and thus make the whole process more efficient. The use of observers in predicting the future of complex systems is not a serious limitation. If weather prediction comes in mind as the prototypical complex system, the observers provide real time sensor information that helps substantially in predictability. Of course, one may investigate whether the introduction of time lags in the information provided by the observers affects dramatically the predictability.

12.6 Summary

- Electron flow in graphene on substrates generates branching.
- Machine learning with observers captures and predicts random singular flow.

References

1. M.A. Topinka, B.J. LeRoy, R.M. Westervelt, S.E.J. Shaw, R. Fleischmann, E.J. Heller, et al., Coherent branched flow in a two-dimensional electron gas. Nature **410**, 183 (2001)
2. H. Degueldre, J.J. Metzger, T. Geisel, R. Fleischmann, Random focusing of tsunami waves. Nat. Phys. **12**, 259 (2016)
3. M. Mattheakis, G.P. Tsironis, Extreme waves and branching flows in optical media, in *Quodons in Mica: Nonlinear Localized Travelling Excitations in Crystals*, eds. by J.F.R. Archilla, N. Jimènez, V.J. Sànchez-Morcillo, L.M. Garcìa-Raffi, Springer Series in Materials Science, pp. 425–454 (2015)
4. M. Mattheakis, G.P. Tsironis, E. Kaxiras, Emergence and dynamical properties of stochastic branching in the electronic flows of disordered Dirac solids. Eur. Phys. Lett. **122**, 27003 (2018)
5. A. Patsyk, U. Sivan, M. Segev, M.A. Bandres, Observation of branched flow of light. Nature **583**(7814), 60–65 (2020)
6. G. Neofotistos, M. Mattheakis, G.D. Barmparis, J. Hizanidis, G.P. Tsironis, E. Kaxiras, Machine learning with observers predicts complex spatiotemporal behavior. Front. Phys. **7**, 24 (2019)

Chapter 13
Discrete Breathers

Intrinsic Nonlinear Localization with Machine Learning

Abstract Discrete breathers are localized nonlinear modes that appear in discrete lattice systems. They can be generated numerically with very precise techniques that also allow the study of their stability and mobility. The use of machine learning techniques permits learning and subsequent recognition of discrete modes through convolutional neural networks. The latter can learn and then differentiate discrete breathers from linearized phonon modes. The breathers are localized in space and time periodic solutions of nonlinear discrete lattices, while phonons are the linear collective oscillations of interacting atoms and molecules. Deep learning neural networks are seen to be able not only to distinguish breather from phonon modes but also to determine with high accuracy underlying nonlinear on-site potentials that generate breathers. These results mean that it is indeed possible to use neural networks and solve inverse nonlinear problems.

13.1 Introduction

Discrete breathers (DBs) or Intrinsic Localized Modes (ILMs) are time periodic and space localized modes that appear in discrete nonlinear lattices [1, 2]. During the several decades there has been intense theoretical and experimental activity that generated a body of precise knowledge regarding these modes [3]. In the present chapter we give a brief description on the physics and mathematics of discrete breathers before we dive into the tools of machine learning that will enable us to characterize them. The basic nature of discrete breathers is that they are periodic in time, much like phonons are in ordered solids. However, unlike phonons, they are localized in space even if there is no disorder in the system. The source of their localization stems from the presence of nonlinearity. This may arise from strong coupling with other degrees of freedom, or it is a genuine property of the lattice system.

We can have nonlinear localization both in Hamiltonian and in open driven systems. In the latter the external drive overcomes dissipation and produces a balanced localized breather state. Breathers of this type may appear in metamaterials, i.e., mesoscopic extended periodic systems that have inherent nonlinear properties.

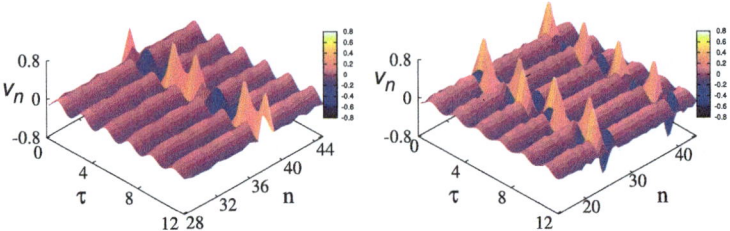

Fig. 13.1 Spatiotemporal evolution of dissipative discrete breathers excited spontaneously in weakly disordered SQUID metamaterials during six periods of the driving flux field. The left and right panels correspond to different configurations of disorder. On the left we see one breather mode while on the right two in different locations of the lattice

In Chap. 11 we gave a brief description of SQUID (Superconducting Quantum Interference Device) metamaterials; they are made of highly nonlinear *oscillators* that are coupled together to the metamaterial [4]. In Fig. 13.1 we show the picture of a one dimensional metamaterial that is driven externally. We notice the periodic oscillation across the lattice; this is the coherent response of the lattice to the driver. We also notice a number of "outliers," i.e., modes that are localized in certain lattice locations and that oscillate with a much higher amplitude. These are dissipative breather modes that are supported by the external, periodic input of energy. These discrete breathers are periodic in time with a high peak in amplitude and smaller amplitude spread in a number of adjacent sites. Since these modes are rather peculiar and interesting, we should like to be able to recognize them and also recognize the details of the systems they are coming from. For this reason we will apply machine learning. Specifically, we would like to know if it is possible to use AI in order to differentiate DBs from phonon modes in simple, Hamiltonian, one dimensional chains with nonlinear on-site potentials. Furthermore it would be very interesting if we could actually tell what systems these breathers are coming from. For the analysis of this inverse mathematical problem, where we know the answer (the breather), but we do not know the question (the lattice) we will use Convolutional Neural Networks (CNNs). This is because the latter are very good in recognizing pictures, and we may use breathers in a way similar to pictures. Since the analysis is completely theoretical, we need to have a precise means for generating breathers numerically. For this purpose we use the numerically exact method introduced by Aubry for generation of DBs from the anticontinuous limit [2].

13.2 Discrete Breathers and Phonons

The model system of interest is a lattice of nonlinear oscillators that are coupled together through certain short-range interaction. We focus on Hamiltonian systems where there is no external driver or dissipation. We consider a one dimensional lattice of N oscillators with a Hamiltonian:

13.2 Discrete Breathers and Phonons

Table 13.1 The three nonlinear on-site potentials used for this study are the hard-ϕ^4, the Morse, and double well potentials

Potentials	
Hard ϕ^4	$V(x) = \frac{x^2}{2} + \frac{x^4}{4}$
Morse	$V(x) = -\frac{1}{2}(1 - e^{-x})^2$
Double well	$V(x) = -\frac{(x-1)^2}{2} + \frac{(x-1)^4}{4}$

$$H = \sum_{n=1}^{N} \frac{p_n^2}{2} + V(x_n) + W(x_n, x_{n+1}), \quad (13.1)$$

where x_n is the displacement of the n-th oscillator from the equilibrium position, p_n is the corresponding momentum, $V(x_n)$ is a nonlinear on-site potential, while $W(x_n, x_{n+1})$ is the nearest neighbor interaction potential. For the latter we assume a harmonic nearest neighbor interaction, viz. $W(x_n) = k/2(x_{n-1} - x_{n+1})^2$ with a coupling parameter k. The value of this parameter is quite important since it determines the existence and precise shape of the discrete breathers of the lattice. The equations of motion for the n-th oscillator are

$$m\frac{d^2 x_n}{dt^2} + \frac{dV(x_n)}{dx_n} = k(x_{n-1} + x_{n+1} - 2x_n), \quad (13.2)$$

where m is the mass of the oscillator that will be taken equal to 1, viz. $m = 1$. For the on-site potential $V(x_n)$, we use three different functions presented in Table 13.1.

The "hard" ϕ^4 potential is a local nonlinear potential with an oscillation frequency that increases as a function of the local oscillator energy. The potential is bounded at infinity. The Morse potential that is very important in Chemical physics has, on the other hand, a frequency that is decreasing as a function of local energy, i.e., is a "soft potential." Furthermore it is unbounded at infinity. Finally the double well potential has a quasi-linear behavior around each of the minima while at larger energies is highly nonlinear since it involves transitions between the two minima. We note that there are no additional parameters in the potentials we chose to investigate; this is in order to simplify the investigation. Thus, once the on-site potential is fixed, the only parameter in the mathematical problem is the coupling strength k of adjacent oscillators.

13.2.1 Discrete Breathers

There are many ways to obtain discrete breather solutions for the lattice systems we focus on [1, 2, 5]. The method introduced by Aubry [2] is an iterative numerically exact procedure that not only produces the breather mode but also determines its stability. The procedure depends critically on the coupling strength k. The iteration starts for the fully decoupled lattice at $k = 0$. In this limit all oscillators are independent of each other. We then select a specific oscillator and construct a *trivial breather mode*, i.e., a nonlinear oscillation in this specific site but no other oscillatory excitations in other sites. Since the oscillator is nonlinear, we may chose the initial excitation energy so that the oscillation proceeds with period T_b or frequency $\omega_b = 2\pi/T_b$. Subsequently, we couple all oscillators together by increasing the coupling k to a small value δk, i.e., $k = \delta k$. This is a crucial step in the procedure since we extend this way the oscillation localized trivially at one site in the whole lattice. However, since the solution is periodic with period T_b, we search for this periodic solution in the whole lattice. Since the coupling k is small, the resulting complex function spanning the lattice is not far from the original trivial breather. Once this solution is found, the procedure repeats itself, and we extend the breather for $k = 2\delta k$, $3\delta k$, ... to the value of coupling desired. While this incremental procedure is followed and we find a breather mode for each coupling value, we also study the stability of the mode by evaluating the tangent map of the transformation [2]. This gives the Floquet spectrum of eigenvalues of the linearized modes around the breather mode; an analysis of this spectrum determines the stability of the mode at all couplings.

Following this numerically exact procedure, we obtain typical breather modes for all three potentials shown in Fig. 13.2. In the upper part of the figure we show the breather mode at a given instant of time while in the lower part the full time evolution for a certain time period. We see that all three breathers are localized in the neighborhood of a lattice site, but their shape and evolution are not identical. Substantial differences in the decay of the amplitude and the morphology of each breather exist.

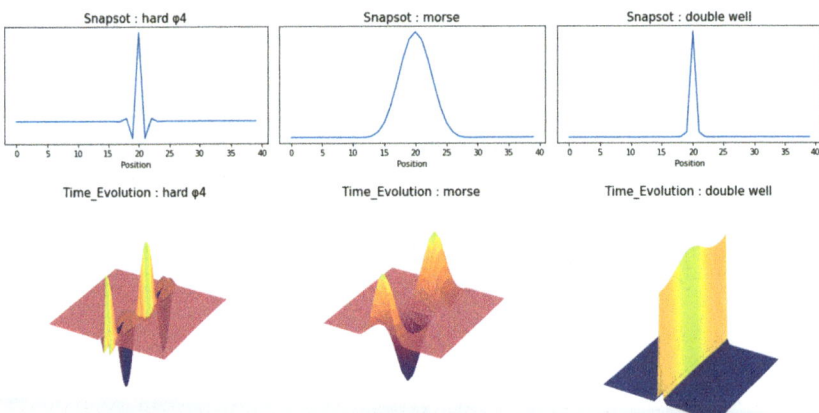

Fig. 13.2 Breather modes at a specific instant of time (upper part) and their time evolution (lower part) for three distinct on-site potentials, viz. hard ϕ^4, Morse, and double well potential. The modes are generated through the application of the procedure from the anticontinuous limit. The value of the coupling parameter is set with coupling $k = 0.1$, while the individual breather frequencies are $\omega_b = 1.317$, 0.967, 0.949, respectively

13.2 Discrete Breathers and Phonons

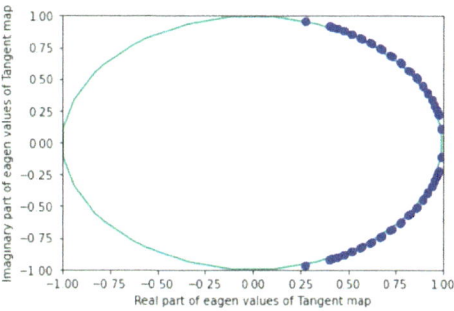

Fig. 13.3 Eigenvalues of tangent map matrix of a breather with potential hard ϕ^4, frequency $\omega_b = 1.227$, and coupling $k = 0.1$. All eigenvalues fall on the (visually distorted) unit circle, and thus the breather is linearly stable

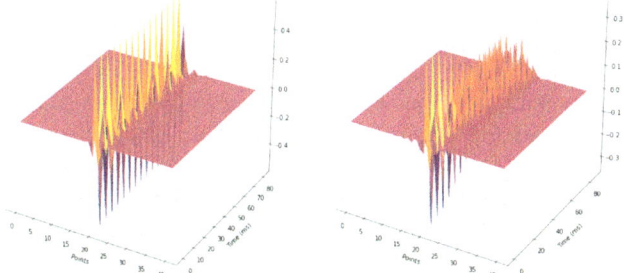

Fig. 13.4 Images of the time evolution over 15 time periods of breathers of the hard ϕ^4 potential with coupling $k = 0.05$. A stable breather with frequency $\omega_b = 1.099$ (left) and an unstable breather with frequency $\omega_b = 1.17$ (right). We notice the decay in the amplitude of the unstable breather at longer times

In addition to constructing the discrete breathers it is also important to know if they are stable or not. We may use the numerically exact procedure through Floquet analysis (Fig. 13.3), where we obtain the eigenvalues of the tangent map to the equations of motion [2]. Stable breathers have Floquet eigenvalues that lie on the circle of radius $r = 1$ on the complex plane. We can also use a less accurate but direct method that involves the observation of the time evolution of the breather modes at long time. In Fig. 13.4, we present the time evolution of a stable (left) and an unstable (right) breather for the hard ϕ^4 potential.

The discrete breather stability depends primarily on how close the breather frequency ω_b is to the spectrum of phonons of the lattice it is generated in. When ω_b is in the band of linearized phonon modes, then the breather is unstable since it resonates with extended lattice modes and thus loses its spectral isolation. The same is true, although this is a higher order effect, when multiples of the breather frequency resonate with phonon modes. Thus, if we wish to have stable breathers, we need to avoid spectral overlap of the breather frequency and its sidebands with phonon modes. If there is such an overlap, there is energy transfer between the local and the extended lattice modes, and the breather loses stability.

13.2.2 Linearized Phonon Modes

Since the aim of this chapter is to show how localized and extended modes differ, we turn now to the phonon modes for the three potentials considered. To obtain the modes we simply linearize the nonlinear potentials keeping only the lowest nontrivial term in each one. For instance, in the ϕ^4 potential we eliminate completely the quartic term x^4 leaving a phonon spectrum of the form

$$\omega(k) \approx \omega_0 + 2k\sin^2\frac{q}{2}. \tag{13.3}$$

where $\omega_0 = 1$, q is the lattice wave-vector, and we considered the coupling k small in order to make a Taylor expansion.[1]

Since we use on-site potentials, we will have an optical frequency with some dispersion in all three cases. Once the phonon spectrum is known, we may obtain easily the time evolution of phonons for each potential of interest. In Fig. 13.5 we show pictures of extended lattice phonons for the hard ϕ^4, Morse, and double well potentials with coupling $k = 0.1$. We used the frequencies $\omega = 1.042, 1.001, 1.415$ for each lattice separately using the criterion that this frequency lies in the phonon band.

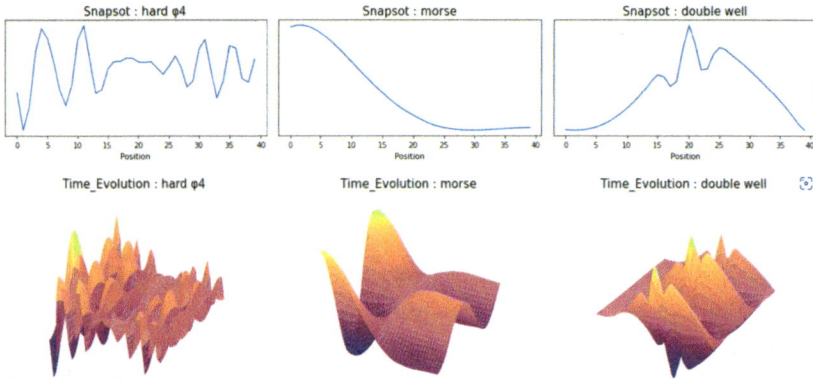

Fig. 13.5 Pictures of phonon modes at a specific time instance (upper) and their time evolution (lower) for the potentials hard ϕ^4, Morse, and double well with coupling $k = 0.1$ and frequencies $\omega = 1.042, 1.001, 1.415$, respectively. In this simulation we used fixed boundary conditions. We notice the extended nature of the phonon modes in contrast to the breather modes that are localized

---

[1] In order to obtain the phonon spectrum we perform a discrete Fourier transform to the linearized equations of motion and use Fourier modes of the form $e^{i(qn-\omega t)}$.

13.2.3 Creation of Breather and Phonon Modes

For the machine learning analysis in this work we create 459 samples of breathers and phonons. For the former we use the anticontinuous limit method in the one dimensional Hamiltonian lattices with the three different nonlinear on-site potentials as outlined in Table 13.1. The DBs have frequencies outside the phonon spectrum, and their stability is controlled through Floquet analysis. We give some details on the methods used below.

13.3 Machine Learning for Breathers and Phonons

Following the short introduction on the physics of breathers and phonons, we now focus on the use of machine learning [6]. The aim is first to use AI in order to differentiate between breathers and phonons and once this is accomplished to also be able to find "microscopic" information on the breathers. We use the numerical procedures described previously and form a dataset of 459 samples that contains breathers and phonons in a balanced way. Each sample constitutes of a single gray-scaled channel of a two dimensional image with size 40×822 pixels that contains the time evolution of discrete breathers or phonons. Each image is constructed in such a way that evolution of at least one time period is included. In Fig. 13.6 we present a colored version of a sample from each category, i.e., a breather and a phonon mode. The complete dataset was shuffled sufficiently in order to avoid biases. The testing

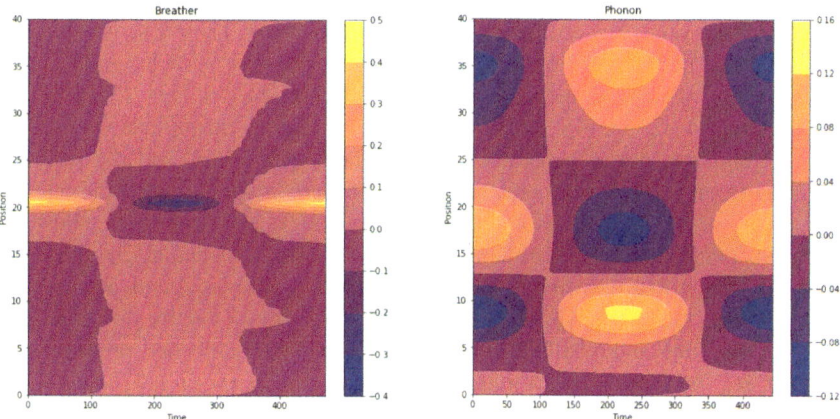

Fig. 13.6 Contour plots of the time evolution of a breather with frequency $\omega_b = 1.332$ and a phonon with frequency $\omega = 1.4161$. The model is the double well potential with coupling $k = 0.1$. The x-axis represents the time, the y-axis the position of each node of the one dimensional chain, while the colorbar the local amplitude. We observe the localized nature of the breather in contrast to the extended trace of the phonon

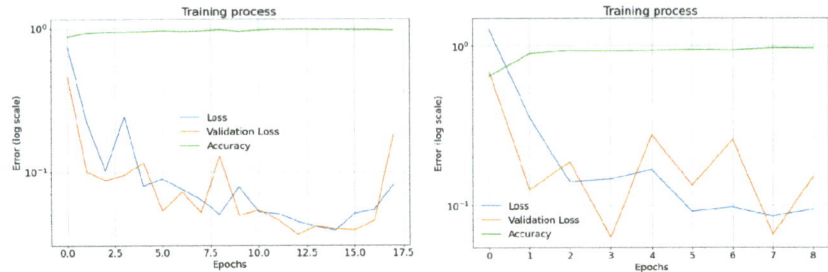

Fig. 13.7 The training accuracy and loss and the validation loss as a function of the number of epochs, for the breather-phonon classification (left) and the potential classification (right) model, respectively

was done on the 20% of the dataset that was separated from the rest of the basis. The remaining 80% of the samples was further split in order to create a training (80%) and a validation (20%) set.

The dataset was normalized using the ImageDataGenerator package [7]. Two models of identical CNNs were constructed. The feature extractor part of each model consisted of three convolutional layers with 32, 64, and 64 ($3x3$) kernels, respectively, and a *ReLU* activation function. The first two convolutional layers in each model were followed by a ($2x2$) Max-pooling layer. The classifier of each model contains two fully connected layers with 64 and 2 nodes, respectively, for the case of the breather/phonon classifier, while and 64 and 3 node layers for the three nonlinear potential classification. A *ReLU* activation function was used for the first layer and a *Softmax* for the output layer for each classifier. The models were allowed to train for 100 epochs, while an early stopping criterion with patience 5 epochs was monitoring the validation error in order to avoid overfitting. The $F1$-score was used to evaluate the performance of the models.

The results of the machine learning analysis are presented in Figs. 13.7 and 13.8 and in Table 13.2. In Fig. 13.7 we show the training quantification data for the differentiation between breathers and phonons (left panel) and the potential detection from the breather modes (right panel). Early stopping limits the number of epochs; we notice the reasonable level of accuracy. In Fig. 13.8 we show the confusion matrix for both breather/phonon and the three potential classification. Both confusion matrices are essentially diagonal signifying good accuracy in both detection processes. This aspect is quantified by the $F1$-score that combines properly precision and recall. The $F1$-scores for both classification processes are high, a fact compatible with the high degree of classification accuracy shown by the confusion matrix.

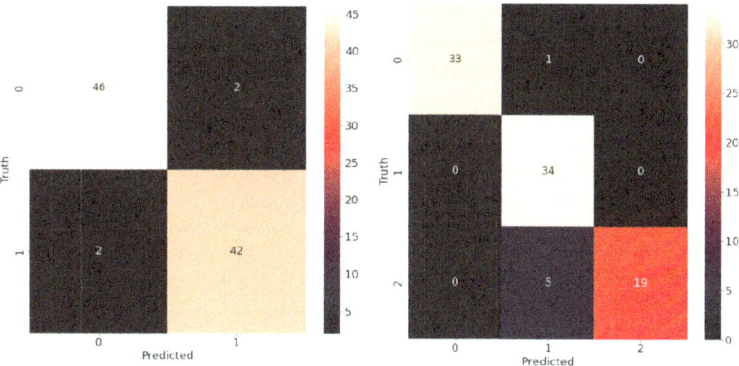

Fig. 13.8 The confusion matrix of the holdout test set for the breather-phonon classification (left) and the potential classification (right) model, respectively. In the left plot "0" indicates phonons, while "1" indicates breathers. In the right plot "0" indicates Morse, "1" indicates double well, and "2" indicates hard ϕ^4 potential, respectively

Table 13.2 Table for the accuracy and the F1-score of the test set for each model

Test set—classification results		
	Breather-phonon	Potentials
Accuracy	95.7%	93.5%
F1-score	0.955	0.934

13.4 Conclusion

In this chapter we presented a simple use of machine learning in differentiating dynamically localized modes as discrete breather with phonon-type extended modes. We generated different lattice modes and used machine learning in order to separate them. Although the results have a limited scope, they nevertheless are quite encouraging in that machine learning can give strong, quantitative insights on the microscopic aspects of nonlinear lattices. We used specifically numerically exact breather and simpler phonon modes in order to train convolutional neural networks that in turn were able to differentiate the modes. We note that similar results were also found through the use of different machine learning methods in the breather problem [8].

We found that it is indeed possible to perform an accurate classification and detect nonlinear breather modes. The quantitative result of the confusion matrix on the test set shows that the classification is excellent provided the CNNs are trained appropriately. This result, however, is not unexpected since breathers and phonons as two dimensional images can be easily distinguished by a human eye. What is more noteworthy, however, is the possibility to find the underlying potential these modes stem from. Indeed, we see that with proper training the machine learning model may distinguish the specifics of the underlying dynamics. This feature opens up the possibility for a more widespread application of deep learning in nonlinear physics. We would very much like to be able to extend this possibility to genuine experimental data and from it to infer the precise underlying dynamics they arise from. More specifically, one could train a machine learning model with a large set of theoretical data and then apply this model to experimental data believed to be compatible with the theoretical ones. The success of such an approach in nonlinear complex systems will help very much the further understanding of complexity through artificial intelligence.

13.5 Summary

- Discrete breathers are space localized and time periodic modes in nonlinear lattice models.
- Phonons are extended modes in lattice systems.
- Convolutional neural networks may differentiate between breathers and phonons and also different breather modes.
- Machine learning can be used in the formulation of the inverse problem of microscopic detection from macroscopic knowledge.

References

1. A.J. Sievers, S. Takeno, Intrinsic localized modes in anharmonic crystals. Phys. Rev. Lett. **61**(8), 970 (1988)
2. S. Aubry, Breathers in nonlinear lattices: Existence, linear stability and quantization. Phys. D **103**, 201 (1997)
3. S. Flach, C.R. Willis, Discrete breathers. Phys. Rep. **295**, 181 (1998)
4. N. Lazarides, G.P. Tsironis, Superconducting metamaterials. Phys. Rep. **752**, 1 (2018)
5. G.P. Tsironis, An algebraic approach to discrete breather construction. J. Phys. A **35**, 951 (2002)
6. T. Dogkas, M. Eleftheriou, D.G. Barmparis, G.P. Tsironis, Identifying discrete breathers using convolutional neural networks, in *Chaos, Fractals and Complexity*, ed. by T. Bountis (Springer, 2023)
7. T. Hastie, R. Tibshirani, J. Friedman, *The Elements of Statistical Learning: Data Mining, Inference, and Prediction*, 2nd edn. (Springer, 2009)
8. J. Bajars, F. Kozirevs, Data-driven intrinsic localized mode detection and classification in one-dimensional crystal lattice model. Phys. Lett. A **436**, 128071 (2022)

Part V
Quantum Complexity

Chapter 14
Quantum Targeted Transfer with Machine Learning

Machine Learning for Quantum Processes

Abstract The classical targeted energy transfer involves a nonlinear resonance that enables efficient transfer from a donor to an acceptor molecule. In the quantum regime a number of bosons may be transferred similarly from a certain crystal site to an alternative one using the same nonlinear resonance. While the process is complex, the use of machine learning enables learning of the optimal quantum paths in the analytically solvable case of a dimer. This knowledge may be extended to more complex quantum walks in configurations that cannot be handled analytically. This application shows that the use of machine learning derived methods in quantum processes enables the efficient design of configurations that lead to efficient quantum devices.

14.1 Introduction

In Chap. 9 we addressed the problem resonant exciton or electron transfer through the targeted energy transfer or TET mechanism [1]. This involves a nonlinear resonance that makes energy or charge transfer very efficient. We used the Discrete Nonlinear Schrödinger (DNLS) equation to formulate it and found the analytical condition for transfer in the case of the dimer. This solution was recaptured through the use of a machine learning approach. The latter enabled the extension of the solution to nonanalytically solvable cases, such as a trimer configuration. Specifically it was shown that minimization of a proper loss function may recover easily the analytically known TET condition [2]. In the present chapter we address a similar question but for the corresponding Quantum TET (QTET) problem that is generated through the quantization of the classical TET problem [3, 4]. This is an interesting extension since it introduces the use of machine learning motivated approaches in a genuine quantum system that has a clear classical counterpart.

14.2 From TET to QTET

We start by expressing the classical TET regime in a Hamiltonian form with notation slightly different from the one in Chap. 9:

$$H = \sum_{k=1}^{f} \omega_k |\psi_k|^2 + \frac{1}{2}\chi_k |\psi_k|^4 - \lambda(\psi_{k+1}^* \psi_k + \psi_{k+1} \psi_k^*), \tag{14.1}$$

where ψ_k, ω_k, and χ_k denote the amplitude, frequency, and the nonlinearity parameter, respectively, of the oscillator at site k, while the parameter λ is the uniform coupling between neighboring sites. In order to quantize the Hamiltonian of Eq. (14.1), we substitute the amplitudes ψ_k^* and ψ_k with the creation and annihilation operators $\hat{a}_k^\dagger$ and $\hat{a}_k$, respectively, as expressed in the second quantization formalism [5]. These operators obey to the boson commutation relations $[\hat{a}_k, \hat{a}_m^\dagger] = \delta_{k,m}$ and $[\hat{a}_k, \hat{a}_m] = 0$, where $\delta_{k,m}$ is the Kronecker delta. Therefore, the Hamiltonian operator in the quantum case becomes

$$\hat{H} = \sum_{k=1}^{f} \left[\omega_k \hat{N}_k + \frac{1}{2}\chi_k \hat{N}_k^2 - \lambda(\hat{a}_{k+1}^\dagger \hat{a}_k + \hat{a}_{k+1} \hat{a}_k^\dagger) \right], \tag{14.2}$$

where $\hat{N}_k = \hat{a}_k^\dagger \hat{a}_k$ is the boson number operator for the site k. Assuming periodic boundary conditions, Eq. (14.2) describes a ring with a discrete number of cells and a fixed total number of bosons. The Hilbert space $\mathcal{H}_N$ related to this problem is finite dimensional, and its dimension coincides with the number of ways one can distribute N *indistinguishable* bosons in f *distinguishable* sites with repetitions. Thus the dimension is

$$\mathcal{D} = \frac{(N+f-1)!}{N!(f-1)!}. \tag{14.3}$$

We use as basis Fock states $|n\rangle \equiv |n_1, n_2, \ldots, n_f\rangle$, where $n_1, n_2, \ldots, n_f$ is the number of bosons at each respective site $1, 2, \ldots, f$ at the state indexed as n. The occupation numbers $\{n_i\}$ satisfy the sum rule $\sum_i n_i = N$. Additionally, the Fock states are orthonormal, i.e.,

$$\langle n | m \rangle = \delta_{n_1, m_1}, \ldots, \delta_{n_f, m_f}. \tag{14.4}$$

The actions of the operators $\hat{a}_k$, $\hat{a}_k^\dagger$, $\hat{N}_k$ on each component $|n\rangle$ of the basis are described by

$$\hat{a}_k |\ldots, n_k, \ldots\rangle = \sqrt{n_k} |\ldots, n_k - 1, \ldots\rangle \tag{14.5a}$$

$$\hat{a}_k^\dagger |\ldots, n_k, \ldots\rangle = \sqrt{n_k+1} |\ldots, n_k+1, \ldots\rangle \tag{14.5b}$$

$$\hat{N}_k |\ldots, n_k, \ldots\rangle = n_k |\ldots, n_k, \ldots\rangle. \tag{14.5c}$$

The dimer case we are initially interested in occurs for $f=2$. As in Chap. 9 we assume that the lower energy state is the *donor* site where nonlinearity is χ_D, while the higher energy site is the *acceptor* one with corresponding nonlinearity χ_A [3].

14.3 Determination of the Loss Function

In the classical TET model we need to find the appropriate values of the two nonlinearity parameters in the donor and acceptor molecules, respectively, so that optimal transfer occurs. Similarly, in the quantum case we search the values of the optimal nonlinearity parameters χ_A, χ_D, so that complete transfer of N bosons occurs between the donor and the acceptor. We assume that all bosons are initially in the donor, and thus we search for the conditions for complete transfer to the acceptor. We may thus employ an algorithm relying on physics informed machine learning that is based on the minimization of a physically motivated loss function [2, 6]. To this effect we need a numerical algorithm for calculating the matrix elements of the Hamiltonian Eq. (14.2). The main challenge in handling this Hamiltonian is labeling the Fock states. We rank the states in *lexicographic* order and assign indices $1, 2, \ldots, \mathcal{D}$ to each configuration of particles among the sites. For instance, assuming $N=2$ and $f=3$, state 1 corresponds to $|1\rangle = [2, 0, 0]$, state 2 corresponds to $|2\rangle = [1, 1, 0]$, and so on, assigning every state configuration to a distinct index. This specific indexing enables computation of the matrix elements $\hat{H}_{ij}$ similarly to [3].

$$\hat{H}_{ij} = \langle i|\hat{H}|j\rangle \stackrel{(14.2)}{=} T_1 + T_2 \tag{14.6a}$$

$$T_1 \equiv \langle i| \sum_{k=1}^{f} \left[\omega_k \hat{N}_k + \frac{1}{2}\chi_k(\hat{N}_k)^2\right] |j\rangle \tag{14.6b}$$

$$T_2 \equiv -\lambda \langle i| \sum_{k=1}^{f-1} \left[a_k^\dagger a_{k+1} + a_k a_{k+1}^\dagger\right] |j\rangle. \tag{14.6c}$$

Calculating T_1 is relatively simple, since it involves the simple action of the boson number operator on state $|j\rangle$ as shown in Eq. (14.5c). Thus Eq. (14.6b) is equivalent to

$$T_1 = \sum_{k=1}^{f} \left[\omega_k j_k + \frac{1}{2}\chi_k(j_k)^2\right] \delta_{i,j}, \tag{14.7}$$

where j_k stands for the number of bosons on site k for the state $|j\rangle$. The calculation of the second term is more complicated, since it involves the action of the interaction operators $a_k^\dagger a_{k+1}$ and $a_k a_{k+1}^\dagger$. The consecutive action of these operators can be explored by referring to Eqs. (14.5a) and (14.5b). These operators, when acting by alone, either create or destroy a boson at given site, but when acting consecutively they create/destroy a boson at a given site while at the same time destroy/create one at a neighboring site. Thus, their action on, e.g., state $|j\rangle$, conserves the total number of bosons, and the resulting state is going to be another state in $\mathcal{H}_N$. Specifically,

$$\langle i|a_k a_{k+1}^\dagger|j\rangle = \sqrt{j_k(j_{k+1}+1)}\delta_{i,p} = C_k^{(p)}\delta_{i,p} \tag{14.8a}$$

$$\langle i|a_k^\dagger a_{k+1}|j\rangle = \sqrt{(j_k+1)j_{k+1}}\delta_{i,m} = D_k^{(m)}\delta_{i,m}. \tag{14.8b}$$

Creating two new states,

$$|p\rangle = |j_1, \ldots, j_k - 1, j_{k+1} + 1 \ldots, j_f\rangle \tag{14.9a}$$

$$|m\rangle = |j_1, \ldots, j_k + 1, j_{k+1} - 1 \ldots, j_f\rangle. \tag{14.9b}$$

Combining the above yields

$$T_2 = -\lambda \sum_{k=1}^{f-1} \left[C_k^{(p)}\delta_{i,p} + D_k^{(m)}\delta_{i,m} \right]. \tag{14.10}$$

With this approach the Hamiltonian matrix can be expressed explicitly, and the eigenstates and eigenvalues can be calculated. The numerical code is written in Python through the use of TensorFlow.

Any initial condition $|\Psi(0)\rangle$ can be expanded to the basis of the eigenstates $|\psi_i\rangle$

$$|\Psi(0)\rangle = \sum_{i=1}^{\mathcal{D}} C_i |\psi_i\rangle, \quad C_i = \langle \psi_i|\Psi(0)\rangle. \tag{14.11}$$

Also, these eigenstates can be expanded to the basis of the Fock states

$$|\psi_i\rangle = \sum_{j=1}^{\mathcal{D}} b_{j,i}|j\rangle, \quad b_{j,i} = \langle j|\psi_i\rangle. \tag{14.12}$$

Using Eqs. (14.11) and (14.12), we can now express the time evolution of the initial wave package at time step t

$$|\Psi(t)\rangle = e^{-i\hat{H}t}|\Psi(0)\rangle = \sum_{i,j}^{\mathcal{D}} C_i b_{j,i} e^{-iE_i t}|j\rangle, \tag{14.13}$$

14.3 Determination of the Loss Function

where E_i is the i-th eigenvalue and $\mathbf{i}$ is the imaginary unit. The time evolution of the average number of bosons at the site k is given by

$$\langle N_k(t) \rangle = \langle \Psi(t) | \hat{N}_k | \Psi(t) \rangle. \tag{14.14}$$

Combining Eq. (14.13) and Eq. (14.14), $\langle N_k \rangle$ can be assessed

$$\langle N_k(t) \rangle = \sum_{i,j,k}^{\mathcal{D}} j_k \cdot C_k^* \cdot C_i \cdot b_{j,k}^* \cdot b_{j,i} \cdot e^{\mathbf{i}(E_k - E_i)t}. \tag{14.15}$$

We now evolve Eq. (14.15) for the acceptor energy level until a predefined time maxt is reached. During this period, the oscillator of the f-th site (acceptor) has completed a number of complete oscillations. The value of the loss function can be calculated by extracting the maximum value from that data. We thus define the loss function LF as

$$LF = N - max\{\langle N_f(t) \rangle\} = N - max\{\langle N_A(t) \rangle\}. \tag{14.16}$$

We note that the loss function is not an explicit mathematical function but depends on the actual quantum dynamics of the evolution. It is thus important to set a reasonable long time for the latter so that the phenomenon is actually captured. In Fig. 14.1 one can observe the characteristic oscillatory behavior when complete transfer occurs, i.e., the system is at resonance.

While the loss function might appear not to have any explicit connection to the parameters χ_A, χ_D we want to optimize for, we can use TensorFlow GradientTape() to keep track of gradient information in terms of the trainable variables χ_A, χ_D throughout the whole process described in this section, i.e., starting

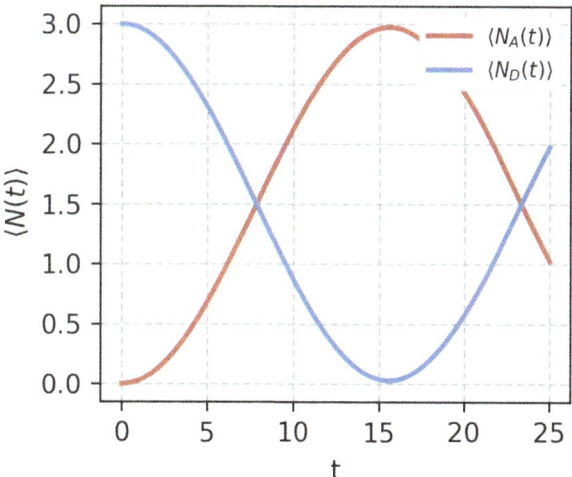

Fig. 14.1 Time evolution of the average number of bosons for the two sites of the dimer under the parameters $(\chi_A, \chi_D, \omega_A, \omega_D, \lambda, N, \mathtt{maxt}) = (-2, 2, 3, -3, 0.1, 3, 25)$

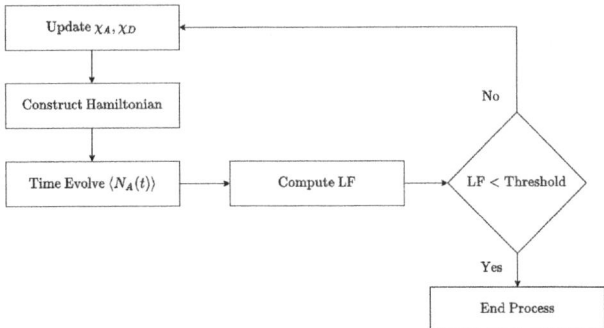

Fig. 14.2 Schematic representation of the parameter optimization procedure

from setting up the Hamiltonian to the calculation of $\langle N_k(t) \rangle$. Using this information we may use an optimizer such as Adam in order to update the parameters in order to minimize the loss function and obtain complete. The parameter `maxt` is of major importance; it controls the specifics of the oscillation of the bosons between the sites. If it is not set large enough, TET can be missed since the system would not have time to complete an oscillation. On the other hand, it has to be as small as possible so that the computation time is minimal.

14.3.1 Optimization Details

The optimization algorithm for the loss function defined proceeds as in Fig. 14.2. The first step of the process is shown in Fig. 14.2, as the update method is the basic part. Since TET is very selective, it is not trivial to optimize the processes when starting with completely random initial parameters. The surface of the loss function we defined exhibits abrupt changes. Its gradient is small throughout the surface, except the area around the optimal parameters where extreme gradients are present. Due to these features of the loss function, we need more precise methods for selecting initial guesses, rather than choosing them randomly.

In order to address these issues we may sharpen the algorithm using one of the following two methods. In the first we may define a grid of initial guesses for the *test optimizers* that will perform a rudimentary initial exploration of the parameter space. Each one of them runs separately, and its iterative procedure is interrupted either due to slow convergence or because of reaching maximum iterations, a parameter defined manually. Subsequently, the *main optimizer* is introduced. Its initial guess is the pair of the parameters corresponding to the minimum loss function derived by the test optimizers. In the second method we may split the parameter space in regions and choose a random combination of initial guesses for each one. The test optimizers iterate as in the first method, and then the main optimizer can be utilized again. If the first run of either method does not produce favorable results, we have the option of redefining the limits of the parameter space around the best parameters provided by the test optimizers. In summary, the first method is versatile and proves to be effective at identifying precisely the parameters leading to TET while being computationally expensive. The second method surmounts this barrier but sacrifices accuracy. In the case of TET, both methods produce accurate results.

14.4 Machine Learning for QTET

After the theoretical analysis presented in the previous section we now turn to the specific machine learning procedures that enable finding the QTET resonances. We will first apply it to the dimer case and then in the unknown case of the trimer.

14.4.1 Quantum Dimer

The results of the application of the loss function optimization methodology detailed in the previous section are shown in Fig. 14.3.

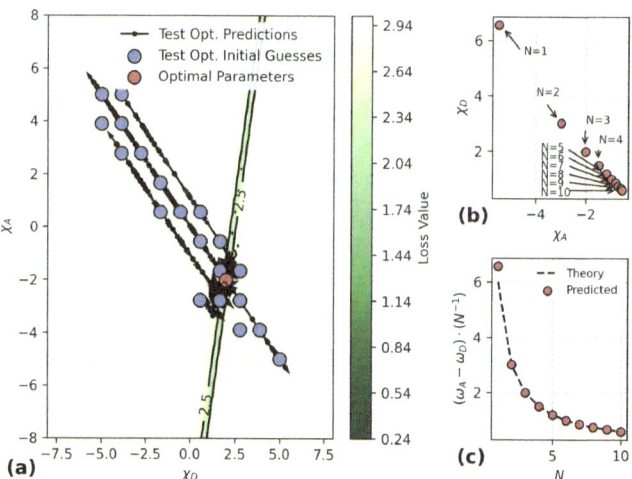

Fig. 14.3 Optimizer results. (**a**) The trajectories and initial guesses of our algorithm using the grid method. The optimal parameters found—by the main optimizer—in this example are $(\chi_A, \chi_D) = (-2.02, 1.99)$ represented as a red dot in the figure. Only the initial guesses of the test optimizers that resulted in a loss value smaller than 0.5 are displayed. (**b**) Parameters for TET from the quantum limit to the classical limit as deduced from the grid method and the main optimizer. (**c**) Comparison of the predicted nonlinearity parameters $\chi_D = -\chi_A$, illustrated in the figure as red dots, with the theoretical value derived from equation, Eq. (14.17), presented in the figure as a dashed line. The system parameters are $(\omega_A, \omega_D, \lambda, N, \texttt{maxt}) = (3, -3, 0.1, 3, 25)$

We observe that test optimizers produce a loss function lower than the threshold mentioned in Fig. 14.2, which is set arbitrarily. The results validate the sensitivity of each optimizer to the initial guesses. Furthermore, the methodology we applied verifies the results known analytically for QTET. It is known that in the classical limit the parameter regime for optimal transfer is [3]

$$\chi_D = -\chi_A = \frac{\omega_A - \omega_D}{N}. \tag{14.17}$$

When we apply the grid method for different number of bosons in the dimer system we produce the results shown in Fig. 14.3b. While Figs. 14.3b and c are generated through the grid method, the same nonlinearity parameters are produced using the method of splitting the parameter space. Indeed, under the parameters of Eq. (14.17), the detuning function, is defined as the variation of the energy of the oscillators during a transfer,

$$\epsilon = [H_D(N) + H_A(0)] - [H_D(i) + H_D(N-i)] \tag{14.18}$$

vanishes. Thus, any nonzero coupling λ can raise the degeneracy of the system. Under these nonlinear parameters, the transfer in the quantum case appears to be most efficient. It is important to point out that higher values of the coupling parameter do lead to a wider range of applicable parameters, which is expected. Maniadis et al. in [3] note that classical and quantum TET is possible even if the detuning function does not vanish. The condition is that λ is greater than a critical value $\lambda_{cr} = \frac{1}{8} I_T |\chi_D + \chi_A|$, where I_T is defined as the total action of the system in the classical case and its expression is $I_T = 2\frac{\omega_D - \omega_A}{\chi_A - \chi_D}$. Moreover, the case of one boson seems to be a special one, since QTET occurs for a whole set of nonlinearity parameters χ_A, χ_D instead of a single one, i.e. a very limited configuration. Specifically we identify this set as

$$\chi_D = \chi_A + 2(\omega_A - \omega_D). \tag{14.19}$$

It can be proven that the detuning function of Eq. (14.18) vanishes for this set of parameters as well. For all the cases where the detuning function vanishes, the Hamiltonian becomes quadratic in the boson operators

$$\hat{H} = E_0 - \lambda(a_D^\dagger a_A + a_A^\dagger a_D). \tag{14.20}$$

Therefore, the time evolution for every quantum observable is identical to the one of the classical case.

14.4.2 Quantum Trimer

While for the TET dimer the resonant transfer condition can be evaluated analytically, the same is not true for larger nonlinear systems. Thus, the machine learning

14.4 Machine Learning for QTET

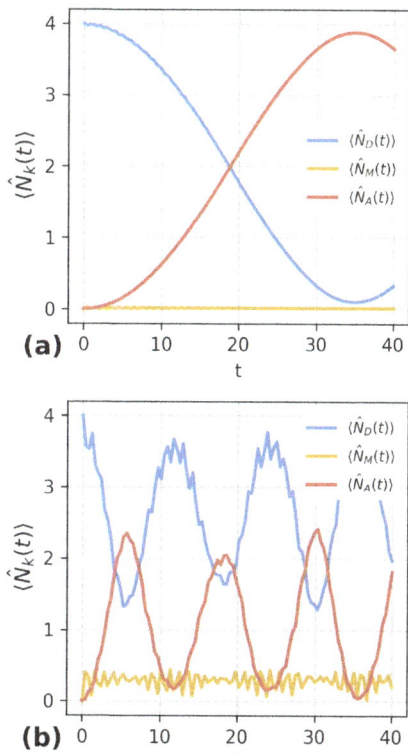

Fig. 14.4 Time evolution of the average number of bosons for the three sites of the trimer system. (**a**) The time evolution of the expectation values of the boson number operators for nonlinearity parameters that produce near complete TET where $(\chi_A, \chi_M, \chi_D, \omega_D, \omega_M, \omega_A, \lambda, N, \text{maxt}) = (-1.5, -38.39, 1.5, -3, -3, 3, 1, 4, 40)$. (**b**) The time evolution of the expectation values of the boson number operators for a system with parameters that don't produce complete TET, where $(\chi_A, \chi_M, \chi_D, \omega_D, \omega_M, \omega_A, \lambda, N, \text{maxt}) = (-1.5, 1.5, 1.5, -3, -3, 3, 1, 4, 40)$. In all of the plots the blue line denotes the donor expectation value, the red one the acceptor and the yellow the middle site

technology we developed for the dimer case and showed that it works there can now become a major tool for the investigation of the transfer in larger systems. The trimer model is simple enough to be the next candidate for the application of the method for arbitrary number of bosons N [2, 3]. In order to simplify the search we set the acceptor and donor sites to their dimer TET/QTET values and optimize the parameter of the intermediate layer that we label as M.

Similarly to the approach in the dimer case we can set an arbitrary threshold for the LF (LF < 0.2) and begin the optimization process. One of the trimer systems examined in this section has the following properties: It is highly coupled, $\lambda = 1$, with frequencies $\omega_D = 3$, $\omega_M = -3$, $\omega_A = -3$ and a maximum number of bosons $N = 4$ initially on the donor site. We observe that the nonlinearity parameter required a value $\chi_M \approx 38.39$, much higher than that of (14.17), compared to the parameters of the donor and acceptor sites $\chi_D = -\chi_A = -1.5$. It is important to mention that the opposite value -38.39 will produce a slightly higher loss function value but still small enough to exceed the threshold and stop the iterative process. We carried out the same procedure for a variety of system parameters, as seen in Fig. 14.4.

14.5 Conclusion

In this chapter we introduced a machine learning method in the context of a quantum many-body system and showed that its efficient implementation can produce results that are very hard to obtain with more conventional methods. We focused on the quantized version of the DNLS equation with arbitrary local energies and nonlinearities and addressed the question of optimal transfer in between different sites. We treated the dimer and trimer cases. Since in the dimer case the fully quantum transfer is known analytically, we compared our method with this case and showed perfect agreement. This successful comparison between analytics and machine learning methods shows that the latter can be used confidently in more complex cases where results are now known. Subsequently we applied the method to the trimer case that cannot be solved analytically. The method enabled a detailed search showing the specifics of the resonant transfer, the different parameter regimes, as well as the transfer efficiencies. In terms of physics we found in the trimer that in the nonresonant transfer from the first (donor) the third (acceptor) sites the intermediate state retains some of the probability. In the resonant case, on the other hand, the intermediate site is essentially non-populated. This shows that this site acts as some form of a barrier between the donor and acceptor sites that can be completely bypassed in the fully resonant regime. It is noteworthy to point out that the bosons move in unison over to the acceptor site showing a very interesting collective behavior in the transfer.

The approach introduced and implemented for the dimer and trimer cases may be extended to collective boson transfer in chains with a larger number of sites. The computational challenge is now larger since the dimensionality of the system becomes large and the calculation of the Hamiltonian and the evolution of $\langle \hat{N}_A(t) \rangle$ is slow. In this regime one needs to explore other loss functions that could improve scalability. The phenomenon of the collective transfer of bosons in the trimer case opens up very interesting new questions on the interplay of nonlinearity and disorder in the fully quantum regime for more extended systems. Machine learning can then be used as a new tool for the investigation of challenging, open problems in condensed matter physics and quantum optics.

14.6 Summary

- Quantum targeted energy transfer works in a similar way as the one for classical variables.
- Machine learning methods can find the nonlinear resonance through appropriate loss function optimization.
- The method is extended and applied in the trimer case where the resonant trajectory is not known.

References

1. G. Kopidakis, S. Aubry, G.P. Tsironis, Targeted energy transfer through discrete breathers in nonlinear systems. Phys. Rev. Lett. **87**, 165501 (2001)
2. G.D. Barmparis, G.P. Tsironis, Discovering nonlinear resonances through physics informed machine learning. J. Opt. Soc. Am. B **38**, C120 (2021)
3. P. Maniadis, G. Kopidakis, S. Aubry, Classical and quantum targeted energy transfer between nonlinear oscillators. Phys. D Nonlinear Phenomena **188**, 153 (2004)
4. I. Andronis, G. Arapantonis, G.D. Barmparis, G.P. Tsironis, Quantum targeted energy transfer through machine learning tools. Phys. Rev. E **107**, 065301 (2023)
5. T.P. Pearsall, *Quantum Photonics*, 2nd edn. Graduate Texts in Physics (Springer International Publishing, Cham, 2020)
6. G.E. Karniadakis, I.G. Kevrekidis, L. Lu, P. Perdikaris, S. Wang, L. Yang, Physics-informed machine learning. Nat. Rev. Phys. **3**, 422 (2021)

Chapter 15
Learning Quantum Systems

Error Reduction with Quantum Autoencoders

Abstract Quantum computing and quantum technologies promise to further revolutionize our world. While in classical information theory the main unit of information is the bit, this is now replaced by the quantum bit or *qubit*. Many of the ideas from machine learning can be transferred to the quantum realm and be used efficiently in quantum computing. In this chapter we give a very brief introduction to quantum neural networks and show how they can be used in order to correct quantum errors that almost unavoidably take place in qubit networks that form quantum computers. We employ as an explicit example quantum autoencoders and use them in a 3-qubit error-correcting framework in order to correct logical qubit states affected by a bit-flip channel. Many important applications of this type will appear in the near future.

15.1 Introduction

Quantum computing is a new frontier in physics, while quantum information science promises a more computational and energy efficient approach to quantitative science over all. It is thus natural to attempt bringing the machine learning approach to the quantum regime [1]. Classical neural networks have demonstrated remarkable capabilities in machine learning, with quantum counterparts holding the promise of handling complex tasks involving quantum instead of classical algorithms. Neural networks and the backpropagation algorithm can identify correlations in data and extract valuable information that may be unattainable through other means. The quantum domain is characterized by two distinct properties, viz. the principle of superposition and that of entanglement. The quantum neural networks may use these principles in order to provide computational advantages over their classical counterparts [2–4]. Moreover, quantum neural networks offer the distinct advantage of processing quantum data without necessitating direct measurements of our quantum systems. This capability makes them especially suited for inherently quantum mechanical processes, such as quantum error correction.

In this chapter we give a flavor on how machine learning may work in the quantum realm. We will explore an implementation introduced for efficient training of the so-called dissipative quantum neural networks (DQNNs); the training is done on data pairs in the form of input and desired output quantum states [5]. This version of quantum neural networks acts as direct analogs of fully connected feed-forward NNs, which trace out qubits from previous layers during the transition to new layers, resulting in energy dissipation; this property gave their name. Work on quantum denoising with quantum autoencoders (QAEs) has showed their ability to denoise specific quantum states [6, 7]. Much like the classical autoencoders, the QAEs have the ability to reconstruct noisy states as well as generate noise-free states. We show examples of this ability by implementing general error-correcting codes using QAEs in arbitrary logical qubit states. Specifically, we will focus on implementing the bit-flip quantum error-correcting code using a quantum autoencoder and demonstrate its ability to correct bit-flip errors in arbitrary logical qubit states [8].

15.2 Quantum Neural Network Architecture

In previous chapters we used a standard, classical approach to neural networks that proved itself very useful in handling complex systems. How useful can be the corresponding quantum networks? The difference between classical and quantum mechanics is that in the latter we have wave like properties in the processes we describe; these quantum waves change the character of the networks into truly probabilistic ones. We start by considering the extension of fully connected feed-forward neural networks in the quantum realm. Our quantum neural networks are fully connected feed-forward networks with a series of consecutive quantum operations. Each layer is connected and acts on the quantum states of the qubits in the next layer. For their training we need a quantum extension of the classical cost function that essentially measures the distance between input and output states.

The quantum networks may be implemented computationally in Python through the package qiskit, assuming that the qubits are noisy-free in general. Initial code may be found in [7]. The code can be altered depending on the details, and one may include optimizers such as RMSprop, Adamax, and Nadam into the already existing ones viz. SGD, Adam, as well as early stopping.[1] Gradient ascent as well as the ability to implement conjugate layers will be discussed below.

We denote by $[m_{in}, m_1, \ldots, m_L, m_{out}]$ a fully connected feed-forward QNN with L hidden layers, each having m_l number of neurons. We can express mathematically the quantum operation for transition from layer $l-1$ to l as in Fig. 15.1 through:

---

[1] *SGD:* Basic gradient descent approach with stochastic updates. Simple but sensitive to the learning rate. *Adam:* Uses adaptive learning rates with momentum, good for noisy and large datasets. *RMSprop:* Adapts learning rates based on the recent magnitudes of gradients, particularly useful for nonstationary data. *Adamax:* Variant of Adam that uses the infinity norm, often more stable for very large datasets and gradients. *Nadam:* Combines Adam with Nesterov momentum for potentially faster convergence.

15.2 Quantum Neural Network Architecture

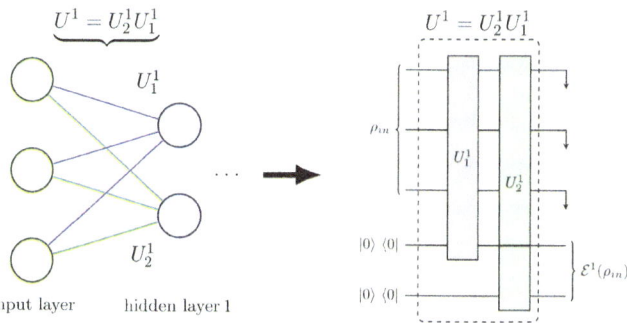

Fig. 15.1 (Left) Input and first hidden layer of a $|3, 2, \ldots|$ quantum neural network depicted schematically where circles correspond to qubits. The transition from the input to hidden layer 1 is done through a unitary transformation U^1, composed of the tensor product of U_2^1 and U_1^1. Each of these unitary operators operates on all of the input qubits of the input layer and a single qubit from hidden layer 1. (Right) Corresponding quantum circuit picture. The initial three qubits (top three lines) denote the input layer, while the subsequent two (bottom two lines) represent hidden layer 1. The operators U_2^1 and U_1^1 operate on these qubits, respectively, while this quantum operation acting on the quantum state of the input layer is labeled as $\mathcal{E}^1(\rho_{in})$. After passing through the circuit, the first three qubits are reset through energy dissipation and then are ready for use in the next quantum layer

$$\mathcal{E}^l\left(\rho^{l-1}\right) \equiv \mathrm{Tr}_{l-1}\left(U^l\left(\rho^{l-1} \otimes |0\ldots0\rangle_l\langle 0\ldots 0|\right)(U^l)^\dagger\right) \qquad (15.1)$$

with $U^l \equiv \prod_{j=1}^{m_l} U_j^l$. In Eq. (15.1), the term ρ^{l-1} represents the quantum state of the qubits in layer $l-1$. The expression $|0\ldots 0\rangle_l\langle 0\ldots 0|$ denotes the quantum state of the qubits in layer l, and, by convention, all these qubits are initialized in the state $|0\rangle$. Subsequently, we apply the unitary transformation U^l to the quantum state $\rho^{l-1} \otimes |0\ldots 0\rangle_l\langle 0\ldots 0|$. We note that the unitary operator, U_j^l, connects all the qubits of the $l-1$ layer with the jth qubit of layer l. The quantum state of the qubits in layer l is therefore obtained by applying, in ascending order of j, all the unitary operators U_j^l and then tracing out all qubits of the previous layer. Then, a QNN simply represents a series of consecutively quantum operations, with its output state being

$$\rho^{out} = \mathcal{E}\left(\rho^{in}\right) = \mathcal{E}^{out}\left(\mathcal{E}^L\left(\ldots \mathcal{E}^2\left(\mathcal{E}^1\left(\rho^{in}\right)\right)\ldots\right)\right). \qquad (15.2)$$

The quantum neural network described just now is capable of simulating a universal quantum computer [9, 10], which effectively means that any quantum algorithm can be built with this method given enough resources. This is certainly one of the most important features of this implementation of quantum neural networks. We need then to perform supervised training in the network and work with the N input/target training pairs of the form $\{|\psi_x^{in}\rangle, |\psi_x^{targ}\rangle\}_{x\in 1,2,\ldots,N}$; for operational simplicity we assume the latter take the form, $|\psi_x^{targ}\rangle = V|\psi_x^{in}\rangle$, where V is an unknown unitary operation that the quantum neural network has to replicate.

In order to evaluate the quantum neural network performance, we define a cost function that returns the distance between the input and output states. One natural choice is to use the fidelity function, viz. $F\left(|\psi_x^{targ}\rangle, \rho_x^{out}\right)$, and average it over all the training pairs, i.e.,

$$C(\kappa) = \frac{1}{N}\sum_{x=1}^{N} \langle \psi_x^{targ}|\rho_x^{out}|\psi_x^{targ}\rangle, \quad (15.3)$$

where κ is a vector that contains all the parameters of the network. The cost function $C(\kappa)$ has two extreme limiting values, i.e., it is 1 when the target and output states are identical or 0 when they are perpendicular to each other. To train the network, we have to maximize the cost function by applying a gradient ascent algorithm or simply minimize $1 - C$ through the more familiar gradient descent algorithm:

$$\kappa^{t+1} = \kappa^t + \eta \nabla_\kappa C(\kappa^t), \quad (15.4)$$

where t denotes the training step or epoch and η the learning rate, typically a small number that ensures that the gradient step is in the vicinity where the cost function decreases.

We parameterize the unitary transformations as follows:

$$U_j^l = e^{iK_j^l} \quad (15.5)$$

with

$$K_j^l = \sum_{\sigma \in P^{\otimes(m_l+1)}} k_\sigma \cdot \sigma, \quad (15.6)$$

where k_σ are real numbers and the parameters to be learned and $P^{\otimes j}$ is the set of all possible tensor products of length j between the elements of $P = \{I, X, Y, Z\}$, i.e., $P^{\otimes 2} = \{II, IX, IY, IZ, XI, XX \ldots\}$. With this parameterization, the U_j^l are uniquely defined through the coefficients k_σ.

The layers l and $l+1$ that have m_l and m_{l+1} number of qubits, respectively, are connected through m_{l+1} number of unitary transformations with 4^{m_l+1} number of coefficients. Thus, the total number of trainable coefficients is

$$\sum_{i=1}^{\ell-1} m_{i+1} \cdot 4^{m_i+1}. \quad (15.7)$$

This number scales exponentially with the number of qubits in a layer. As a result, it becomes unpractical to simulate quantum neural networks containing more than a small number of qubits with classical computers. Additionally, while operating the QNN procedure, in order to construct the quantum map $\mathcal{E}^l$, we reuse qubits from previous layers; this is done by resetting them to the state $|0\rangle$ and enables the use the least amount of qubits necessary. The estimation of the cost function (15.3) is done by implementing the swap test algorithm [11] between the output state of the QNN and the target state that determines their proximity.

In short, by employing a series of consecutive parameterized gates between layers and resetting neurons to $|0\rangle$ state when transitioning to a new layer, we can construct a quantum neural network. Through a supervised training method that utilizes gradient ascent to maximize the fidelity function between the output of the QNN and a target state of our choice, QNNs can effectively learn to perform unknown quantum algorithms.

15.2.1 *Quantum Neural Networks with Conjugate Layers*

In order to implement efficient autoencoder-like quantum neural networks, we use the idea of a conjugate layer constructed by replacing the transformation of a layer, l, with the Hermitian conjugate of the unitary transformation of a different layer of choice, say the l' [8]:

$$U^l_{conj} = (U^{l'})^\dagger = \left(\prod_{j=1}^{m_{l'}} U^{l'}_j\right)^\dagger = \prod_{j=m_{l'}}^{1} (U^{l'}_j)^\dagger, \quad (15.8)$$

where

$$(U^{l'}_j)^\dagger = e^{-iK^{l'}_j}. \quad (15.9)$$

As a result these two layers are trained simultaneously, and thus we essentially truncate the number of trainable parameters of the quantum neural network since the parameters comprising layer l also characterize layer l'.

In order to implement this method efficiently, the conjugate layer must apply the conjugate unitary transformations in reverse order. For instance, if layer l contains m neurons that are transformed into n in the subsequent layer, the conjugate layer should have n neurons that are transformed back into m. As a result, the QNN will exhibit the following structure:

$$[\ldots, \underbrace{m, n}_{U^l}, \ldots, \underbrace{n, m}_{U^l_{conj}}, \ldots].$$

This approach enables the integration of multiple conjugate layers within the quantum neural network architecture, accommodating up to q conjugate layers

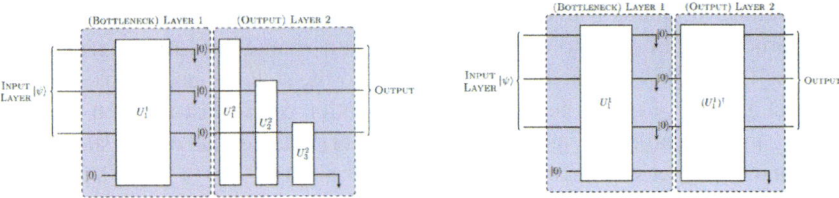

Fig. 15.2 (Left) Vanilla [3,1,3] quantum neural network. (Right) Similarly, a [3,1,3] quantum neural net with a conjugate layer

in a network with a total of $2q + 1$ layers. As an example we show two versions of the same quantum neural network employed for error-correcting bit-flip errors. In Fig. 15.2, the left circuit represents the vanilla quantum neural network implementation, while the right one replaces the transformation of the output layer with the Hermitian conjugate of the first layer. We observe that the three two-qubit unitary transformations, $U_{1,2,3}^2$, are replaced by a single four-qubit transformation $(U_1^1)^\dagger$.

15.3 Quantum Autoencoders for Bit-Flip Error Correction

Classical autoencoders are typically employed for removing unwanted features, e.g., noise, from the dataset. Quantum autoencoders are very much inspired by their classical counterparts and can be set to perform quantum error correction by removing the least relevant features from quantum states, such as quantum noise. Unlike conventional quantum error correction algorithms that typically require generalized measurements and classical information processing, the quantum autoencoders are supposed to be capable of performing those tasks autonomously. In this section, we will investigate if and how quantum autoencoders are able to denoise specific types of errors, by utilizing quantum error-correcting codes. The specific approach we follow is to first fit the quantum autoencoder with multiple different encoded states that are corrupted by a quantum channel, denoted as $|\tilde{\psi}\rangle_L$, and subsequently train the quantum autoencoder to replicate the states with no error, $|\psi\rangle_L$. Our focus is on correcting errors generated by the bit-flip channel that applies a NOT gate to every qubit with probability p; it is described by the following two operation elements:

$$E_0 = \sqrt{1-p}\, I = \sqrt{1-p} \begin{bmatrix} 1 & 0 \\ 0 & 1 \end{bmatrix} \quad E_1 = \sqrt{p}\, X = \sqrt{p} \begin{bmatrix} 0 & 1 \\ 1 & 0 \end{bmatrix}, \quad (15.10)$$

where I is the identity matrix corresponding to the case where the qubit state was left uncorrupted and X is the Pauli matrix σ_x responsible for the bit-flip. We use a 3-qubit error-correcting code and encode all of the input/target states as follows:

$$|\psi\rangle = a|0\rangle + b|1\rangle \rightarrow |\psi\rangle_L = a|000\rangle + b|111\rangle. \quad (15.11)$$

15.3 Quantum Autoencoders for Bit-Flip Error Correction

In order to train the quantum autoencoder to denoise bit-flips, we create a training set of 120 input/target pairs of the form $\{|\tilde{\psi}\rangle_L, |\psi\rangle_L\}$, where $|\psi\rangle_L$ is one of the following states: $|0\rangle_L, |1\rangle_L, |+\rangle_L, |-\rangle_L, |+i\rangle_L, |-i\rangle_L$.[2] The input states of the set are corrupted by (single qubit) bit-flip errors, with probability $p = 0.2$. Additionally, the models were trained in multiple training sessions. Each session had fixed hyperparameters and continued the training from the model of the previous session. Initial sessions generally had a relatively large learning rate ($lr = 0.25$) with Adam or Nadam optimizer and a batch size of 20. Later sessions had decreased learning rate, vanilla SGD optimizer, and no batch size. This strategy was adapted such that we ensure convergence of the cost function as well as to save computational time. Lastly, the initialization of the quantum autoencoder parameters was chosen to be uniformly random between 0 and 1, rather than all being fixed to 0 in order to avoid a plateaued starting point (Fig. 15.3).

To evaluate the model's performance, we create a validation set during training, which ensures that the quantum autoencoders generalize and correct bit-flip errors from any arbitrary state. To this end, we may parameterize the (logical) qubit's state in the Bloch sphere representation

$$|\psi(\theta, \phi)\rangle_L = \cos(\theta/2)|0\rangle_L + e^{i\phi}\sin(\theta/2)|1\rangle_L. \quad (15.12)$$

We may use a meshgrid with $N = 20$ different values for the parameters θ and ϕ, where $\theta \in [0, \pi]$ and $\phi \in [0, 2\pi]$. The validation set, thus, contains in total $400(=$

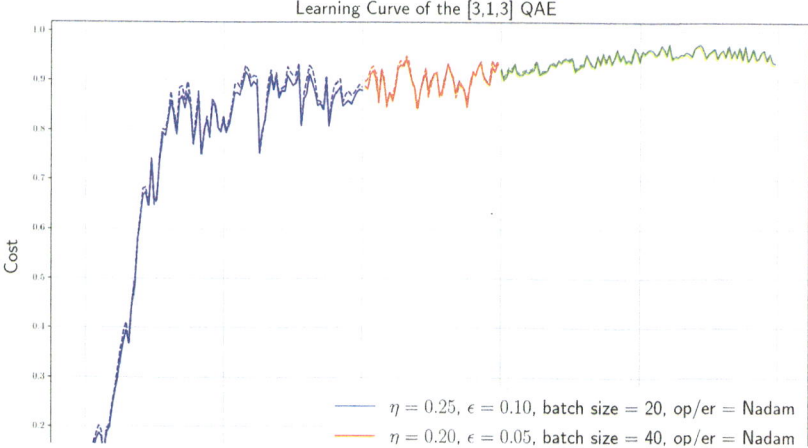

Fig. 15.3 Learning curve of the [3, 1, 3] quantum autoencoder. The dashed line indicates the training set cost function, while the continuous one, on the validation set. Each color represents a learning session with fixed hyperparameters. The training continues for another 250 epochs (500 in total), but the model showed minimal improvement in those last epochs

---

[2] $|+i\rangle = \frac{|0\rangle + i|1\rangle}{\sqrt{2}}$ and $|-i\rangle = \frac{|0\rangle - i|1\rangle}{\sqrt{2}}$.

20 × 20) different states, such that they uniformly cover the whole Bloch sphere. Those states were then corrupted with a bit-flip error with probability $p = 0.2$. The mean fidelity ($\equiv$ cost of the validation set) is calculated by

$$\bar{F} = \frac{1}{4\pi} \int_0^{2\pi} \int_0^{\pi} F(\rho_{out}(\theta, \phi), |\psi(\theta, \phi)\rangle_{targ}) \sin(\theta) d\theta d\phi \qquad (15.13)$$

$$\simeq \frac{1}{4\pi} \sum_{i=0}^{N} \sum_{j=0}^{N} F(\rho_{out}(\theta_i, \phi_j), |\psi(\theta_i, \phi_j)\rangle_{targ}) \sin(\theta_i) \Delta\theta \Delta\phi \qquad (15.14)$$

with $\Delta\theta = \pi/N$, $\Delta\phi = 2\pi/N$, $\theta_i = i\Delta\theta$, and $\phi_j = j\Delta\phi$.

Following this strategy, we train multiple models with different hyperparameters each and make sure there are no overfitting issues. The quantum autoencoder successfully learns to denoise bit-flips, with the cost function on both the train and validation sets eventually reaching a value approximately equal to 0.98 [8]. It would be useful to study the performance of the model given that a bit-flip took place in one of the three qubits that comprise the logical qubit or that no error has occurred. To achieve this, we calculated the corresponding conditional fidelities, on four validation sets

$$\bar{F}(\rho_{out}, |\psi\rangle|\text{bit-flip to qubit 1}) = 0.96$$

$$\bar{F}(\rho_{out}, |\psi\rangle|\text{bit-flip to qubit 2}) = 0.96$$

$$\bar{F}(\rho_{out}, |\psi\rangle|\text{bit-flip to qubit 3}) = 0.95$$

$$\bar{F}(\rho_{out}, |\psi\rangle|\text{no error}) = 0.97.$$

This confirms that the quantum autoencoder has successfully learned to denoise single bit-flip events on (logical) qubits, with high fidelity. We present the colormaps in Fig. 15.4 depicting the fidelity $F(\rho(\theta, \phi), |\psi(\theta, \phi)\rangle)$ as a function of the

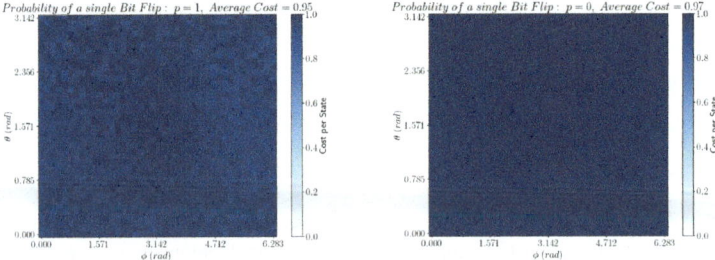

Fig. 15.4 Each point in the panels corresponds to a qubit state on the Bloch sphere (15.12). The autoencoder was tested with $N = 1600$ different qubit states

parameters θ and ϕ on two different validation sets. In the first validation set, the probability of error was set to be $p = 1$, while in the second one $p = 0$.

With this method, we essentially check if the quantum autoencoder has developed a bias during training toward any specific region on the Bloch sphere. The colormaps would reveal this kind of bias if their color was not uniform. In such cases, the training set must be accordingly modified and include a representative state from that region so that the autoencoder can generalize more efficiently. Figure 15.3 suggests that there are no areas on the Bloch sphere where the model underperforms.

While employing preexisting error-correcting codes in quantum autoencoder models can be advantageous, it is important to recognize that these models also inherit the limitations of such methods. For instance, if two or more bit-flip errors occur, the QAE may fail to accurately identify and correct the error, leading to improper error correction for that particular state. In an optimal quantum error-correcting algorithm, the probability of successful correction is given by the expression $3(1 - p)p^2 + p^3$. Although the quantum autoencoder models presented here are not the optimal ones, they nevertheless demonstrate the ability to recover corrupted states with high fidelity. Thus we find that probability of successful correction follows closely the same dependence as similar optimal quantum error-correcting algorithms.

15.4 Conclusions

In this chapter we explored the use of quantum autoencoders and quantum neural networks to design and implement quantum error-correcting codes for various quantum channels. Our investigation focused on a bit-flip channel where we successfully employed quantum autoencoders to error-correcting corrupted states. A deeper exploration into the architecture of the [3,1,3] autoencoder offers intuition on the observed phenomena. Drawing from the relationship $\sum_{i=1}^{\ell-1} m_{i+1} \cdot 4^{m_i+1}$ [7], we can determine the number of trainable parameters in a model with l layers and architecture $[m_1,\ldots,m_i,\ldots,m_l]$. Specifically, a [3,1,3] model encompasses 304 trainable parameters. Autoencoders, with such a significant number of parameters relative to their size, present challenges in optimal training. Often, these models are susceptible to entrapment in local minima or, as in this instance, local maxima of the cost function. The effect of conjugate layers, which effectively reduce the number of trainable parameters, further supports this assertion, as models incorporating them tend to outperform vanilla implementations.

We note that dissipative quantum neural networks can sometimes exhibit plateaus while training, which prevent them from reaching the global maximum of the cost function and consequently make it challenging to train them for specific purposes [8]. This observation is in parallel with the work [12], where the authors characterize such quantum neural networks as untrainable due to the barren plateaus they tend to fall into during training. These facts may limit the possible use of these networks although other approaches may address more efficiently this issue. For instance, in the present exposition we saw that conjugate layers can offer significant help in training these models while also accelerating the training process.

Quantum neural networks have the potential to discover unknown encryptions for unconventional noisy channels, opening up new possibilities in the field of quantum computing. As quantum technology continues to advance, the ability to adapt and optimize error-correcting codes for different quantum channels will become increasingly crucial in ensuring reliable and efficient quantum systems.

15.5 Summary

- Quantum machine learning may be applied in qubit systems for error correction.
- A quantum autoencoder configuration is used to reduce spin flip errors in a three-qubit system.
- Barren plateaus where the loss function becomes very small with the increase of the system size make difficult the generalization of simple methods.

References

1. J. Biamonte, P. Wittek, N. Pancotti, P. Rebentrost, N. Wiebe, S. Lloyd, Quantum machine learning. Nature **549**, 195 (2017)
2. S.A. Stein, R. L'Abbate, W. Mu, Y. Liu, B. Baheri, Y. Mao, G. Qiang, A. Li, B. Fang, A hybrid system for learning classical data in quantum states. arXiv:2012.00256 (2021)
3. D. Uke, K.K. Soni, A. Rasool, Quantum based support vector machine identical to classical model, in *2020 11th International Conference on Computing, Communication and Networking Technologies (ICCCNT)* (IEEE, 2020)
4. J.M. Martinis, et al., Quantum supremacy using a programmable superconducting processor. Nature **574**, 505–510 (2019)
5. K. Beer, D. Bondarenko, T. Farrelly, T.J. Osborne, R. Salzmann, D. Scheiermann, R. Wolf, Training deep quantum neural networks. Nat. Commun. **11**, 808 (2020)
6. D. Bondarenko, P. Feldmann, Quantum autoencoders to denoise quantum data. Phys. Rev. Lett. **124**, 130502 (2020)
7. T. Achache, L. Horesh, J. Smolin, Denoising quantum states with quantum autoencoders—Theory and applications. arXiv:2012.14714 (2020)
8. A. Chalkiadakis, M. Theocharakis, G.D. Barmparis, G.P. Tsironis, Quantum neural networks for the discovery and implementation of quantum error-correcting codes. Chaos **33**, 113127 (2023)
9. D. Deutsch, Quantum theory, the Church–Turing principle and the universal quantum computer. Proc. R. Soc. Lond. A **400**, 97–117 (1985)
10. R.W.K. Beer, D. Bondarenko, Efficient learning for deep quantum neural networks. Nat. Commun. **11**, 808 (2019)
11. A. Barenco, A. Berthiaume, D. Deutsch, A. Ekert, R. Jozsa, C. Macchiavello, Stabilization of quantum computations by symmetrization. SIAM J. Comput. **26**, 1541 (1997)
12. K. Sharma, M. Cerezo, L. Cincio, P.J. Coles, Trainability of dissipative perceptron-based quantum neural networks. Phys. Rev. Lett. **128**, 180505 (2022)

Part VI
Biomedical Applications

Chapter 16
Action Potential Propagation in the Heart

Electricity in the Heart

Abstract The heart is a rather complex organ in humans and animals that works through the coordination of electrical pulses with muscle contraction. It beats due to synchronized contraction of myocytes initiated at the sino-atrial node through electrical action potentials that subsequently propagate through the heart. Cardiac arrhythmias are anomalies such as atrial and ventricular fibrillation, ventricular tachycardia, etc., that are generated through propagation disruption and instabilities in the electrical signals. We develop here general nonlinear models that describe the propagation of the action potentials through the heart tissue. As these signals propagate through the heart, they scan the various tissues that alter the specifics of their evolution in the different locations. The electrocardiogram gives a picture of this pulse propagation. The pseudo-electrocardiogram is a signal generated by the model pulses and contains only partial information compared to the one in a real clinical electrocardiogram.

16.1 Introduction

The heart is a unique organ that is central to life; its main function is to pump blood in the body (Fig. 16.1). It operates electro-mechanically through generation of electrical pulses in the upper right part that in turn propagate through the heart, couple to the cardiac cells, and generate periodic muscle contraction. The electrical cardiac pulse acts as a pump, while, at the same time, is used as an indicator of the function and state of the organ. In the sino-atrial node there are few million pacemaker cells that generate locally an electrical action potential that depolarizes myocytes, the muscle cells of the heart. This electrical depolarization propagates in turn through the linked network of cells connected via gap junctions. The electrical dynamics in muscles was discovered in 1950s by Hodgkin and Huxley who made experiments in the long axon of the neuron of the giant squid. They wrote a concise mathematical equation that describes the process of propagation of the action potential [1]. Their ideas are passed to the heart dynamics through mathematical models that are a bit more specific to the heart. This modeling is very important in the understanding of arrhythmias as well as other heart problems [2].

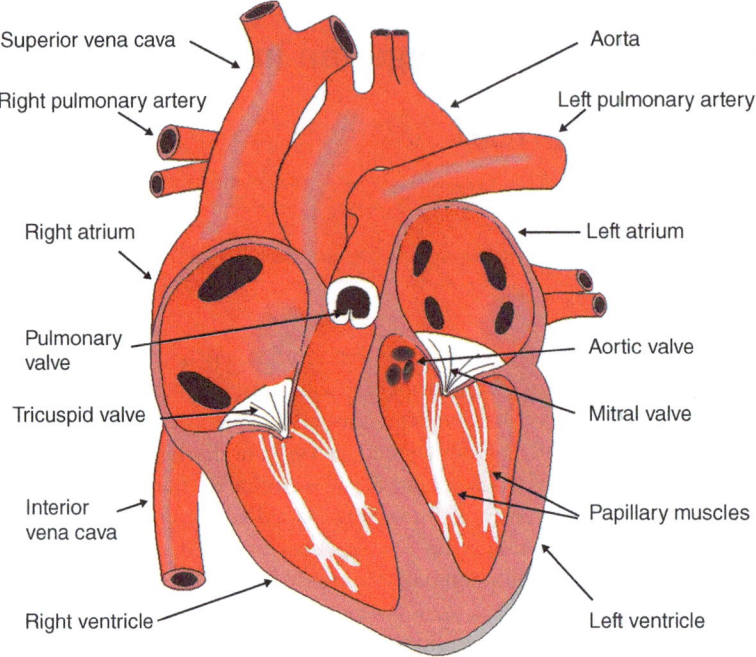

Fig. 16.1 A sketch of the different parts of heart. Reprinted from the free open-access book [7]

From the point of view of the electrical pulse propagation, the heart muscle is a medium with variable conductivity that depends on the local constituent cells as well the local morphology. Electricity is provided by the various ions such as sodium, potassium, calcium, etc., that are gated in and out of the cardiac cells. Following the Nobel award winning work of Hodgkin and Huxley as well as subsequent efforts of numerous scientists, we need to set up a mathematical model that takes into account quantitatively how the ionic currents affect and modify the action potential pulse propagation. The model has to be able to describe features such as the duration of the action potential, propagation speed, amplitude, and finally result in an effective electrocardiogram signal that is reasonable and compatible with the natural one.

Significant information on cardiac dynamics and modeling can be found in recent literature [3–6]. For simplicity here we will focus on a single model that is relatively realistic while containing most of the basic electrical cardiac dynamics. This is a three-variable cardiac model introduced by Fenton and Karma in 1998 that we will be abbreviated as FK model [6]. This FK model is employed for the calculation of the ventricular action potential over a one dimensional cable transversal to the ventricular tissue (Fig. 16.2). For simplicity, only one set of electrophysiological parameters is used, i.e., the cable spans a single region of the ventricular tissue. Using the action potential obtained for external stimulus current of various characteristics, we calculate the corresponding cardiogram signal. This is termed pseudo-electrocardiogram (pECG) since although it contains many of the clinical cardiograph signals it does not provide a full realistic model for it [8–12].

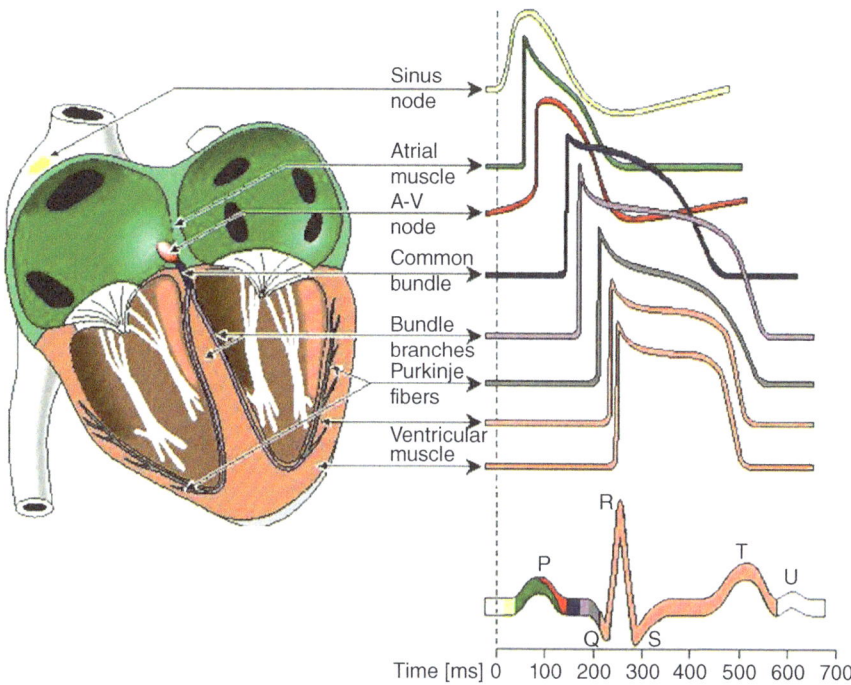

Fig. 16.2 Action potential shapes in the various regions of the human heart, and how they are related to the electrocardiogram. Specifically, the P waves are related to the atrial signal, the QRS complex to the depolarization of the ventricles, and T waves to the repolarization of the ventricles. Image courtesy of Dr. De Voogt and ECGpedia.org

16.2 The Ventricular Fenton-Karma Model

In this section we describe a basic model for the electric pulse propagation in the ventricles of the heart; it was introduced by Fenton and Karma in 1998 and is based on the original work of Hodgkins and Huxley [1, 4, 6]. The ventricular FK model consists of three equations:

$$\frac{\partial u}{\partial t} = \nabla \cdot \left(\tilde{D}\nabla u\right) - J_{fi}(u;v) - J_{so}(u) - J_{si}(u;w), \tag{16.1}$$

$$\frac{\partial v}{\partial t} = \Theta(u_c - u)\frac{1-v}{\tau_v^-(u)} - \Theta(u - u_c)\frac{v}{\tau_v^+}, \tag{16.2}$$

$$\frac{\partial w}{\partial t} = \Theta(u_c - u)\frac{1-w}{\tau_w^-} - \Theta(u - u_c)\frac{w}{\tau_w^+}. \tag{16.3}$$

The first equation is a diffusion-type equation for the dependent variable of the normalized action membrane potential u, while the other two are nonlinear

Table 16.1 Three different sets of model parameters that enter into the Fenton-Karma model, Eqs. (16.1)–(16.3), from Reference [6]. "M" stands for "modified"

Parameter	BR model	MBR model	MLR-I model
$\tilde{g}_{fi}$	4	4	5.8
τ_r	33.33	50	130
τ_{si}	29	44.84	127
τ_0	12.5	8.3	12.5
τ_v^+	3.33	3.33	10
τ_{v1}^-	1250	1000	18.2
τ_{v2}^-	19.6	19.2	18.2
τ_w^+	870	667	1020
τ_w^+	41	11	80
u_c	0.13	0.13	0.13
u_v	0.04	0.055	–
u_c^{si}	0.85	0.85	0.85
Other parameters			
C_m	$1\,\mu\mathrm{F/cm}^2$		
V_0	$-85\,\mathrm{mV}$		
V_{fi}	$+15\,\mathrm{mV}$		
$\tilde{D}$	$0.001\,\mathrm{cm}^2/\mathrm{ms}$		
k	10		

equations for the two gate variables v and w. The normalized action potential is expressed in terms of the true action potential V measured in units of millivolts (mV) as follows:

$$u \equiv \frac{V - V_0}{V_{fi} - V_0}, \qquad (16.4)$$

where V_0 is the resting membrane potential, and V_{fi} is the Nernst potential of the fast inward current. These two parameters are important for setting the scales for the electrical pulse. The reduced variable u varies from 0 to 1. The time parameter $\tau_v^-(u)$ depends on u and is expressed as follows:

$$\tau_v^-(u) = \Theta(u - u_v)\tau_{v1}^- + \Theta(u_v - u)\tau_{v2}^-. \qquad (16.5)$$

This splitting allows us to separate independently the minimum diastolic interval, i.e., the excitable gap, controlled by τ_{v1}^-, and the steepness of this curve, controlled by τ_{v2}^-. The values of the variables V_0, V_{fi}, u_v, τ_{v1}^-, and τ_{v2}^-, along with the values of the other parameters that appear in Eqs. (16.1)–(16.3), i.e., $\tilde{D}$, u_c, τ_v^+, τ_w^-, and τ_w^+, are given in Table 16.1.

The variables J_{fi}, J_{so}, and J_{si} denote the scaled phenomenological ionic currents related to the corresponding true currents in units of mA through the following equations:

16.2 The Ventricular Fenton-Karma Model

$$J_\alpha = \frac{I_\alpha}{C_m(V_\alpha - V_0)}, \quad (16.6)$$

where C_m is the membrane capacitance, and $\alpha = fi$, so, and si, with I_{fi} being a fast inward current that is responsible for depolarization of the membrane and only depends on one inactivation-reactivation gate v. The latter is responsible for inactivation of the current after depolarization and its reactivation after repolarization. The I_{so} is a slow outward current that is analogous to the time-independent potassium current and is responsible for repolarization of the membrane. The I_{si} is a slow inward current, analogous to the calcium current, that balances I_{so} during the plateau phase of the action potential and only depends on one gate variable w. The latter is responsible for inactivation and reactivation of this current. In a classic physiological picture of membrane dynamics, the variables I_{fi}, I_{so}, and I_{si} correspond to the Na, K, and Ca currents, respectively. In general this correspondence presents an oversimplification since the known membrane dynamics is considerably more complex. The explicit expressions for the normalized currents are the following:

$$J_{fi}(u; v) = -\frac{v}{\tau_d}\Theta(u - u_c)(1 - u), \quad (16.7)$$

$$J_{so}(u) = +\frac{u}{\tau_0}\Theta(u_c - u) + \frac{1}{\tau_r}\Theta(u - u_c) \quad (16.8)$$

$$J_{si}(u; w) = -\frac{w}{2\tau_{si}}\left\{1 + \tanh\left[k\left(u - u_c^{si}\right)\right]\right\}, \quad (16.9)$$

where

$$\tau_d = \frac{C_m}{\bar{g}_{fi}}, \quad (16.10)$$

and the values of the parameters $\bar{g}_{fi}$, τ_0, τ_r, τ_{si}, k, and u_c^{si} are also given in Table 16.1. The function $\Theta = \Theta(x)$, which appears repeatedly in Eqs. (16.1)–(16.3) and Eqs. (16.7)–(16.9), is the standard Heaviside step function defined by $\Theta(x) = 1$ for $x \geq 0$ and $\Theta(x) = 0$ for $x < 0$.

It is important to notice that the diffusion coefficient $\tilde{D}$ is a function of spatial coordinates since the tissue in the heart changes from place to place. In simple one dimensional case we will consider here, the first term on the right-hand side of Eq. (16.1), i.e., $\nabla \cdot \left(\tilde{D}\nabla u\right)$ becomes

$$\frac{\partial}{\partial x}\left(\tilde{D}(x)\frac{\partial u(x)}{\partial x}\right) = \frac{\partial \tilde{D}(x)}{\partial x}\frac{\partial u(x,t)}{\partial x} + \tilde{D}(x)\frac{\partial^2 u(x,t)}{\partial x^2}. \quad (16.11)$$

16.3 Numerical Calculation of the Action Potential

Equation (16.1) is of reaction-diffusion type; the diffusive aspect is set by the first order time derivative and second order space derivative terms. The reactive terms are the input currents that depend on the additional two Eqs. (16.2) and (16.3). The set of equations is rather complex, and it is very hard to solve it analytically, at least in its very general form. Since we are dealing with a partial differential equation, we need to set also the boundary conditions. We may choose Neumann boundary conditions that signify that there is no flux at the end of the cable, i.e., the one dimensional propagation path of the pulse:

$$\tilde{D}(x) \left.\frac{\partial u(x,t)}{\partial x}\right|_{x=0} = \tilde{D}(x) \left.\frac{\partial u(x,t)}{\partial x}\right|_{x=L} = 0, \qquad (16.12)$$

where L is the length of the cable. Here, the value of the length L is set everywhere to $L = 3.5$ cm. In order to integrate equations (16.1)–(16.3) with the boundary conditions Eq. (16.12), we discretize the spatial domain into $N_x - 1$ elements with N_x nodes at $x_i = (i-1)L/(N_x - 1)$. These nodes are separated by distance that is $\Delta x = L/(N_x - 1)$ ($i = 1, 2, ..., N_x$).

In order to solve numerically the equations, we need to discretize the derivative terms. From Eq. (16.11) we can see that we need the discrete form of the first and second spatial derivative of $u(x, t)$, as well as the first spatial derivative of $\tilde{D}(x)$. We use the following centered differences:

$$\frac{\partial u(x,t)}{\partial x} = \frac{u_{i+1}(t) - u_{i-1}(t)}{2\Delta x}, \qquad (16.13)$$

$$\frac{\tilde{D}(x)}{\partial x} = \frac{\tilde{D}_{i+1} - \tilde{D}_{i-1}}{2\Delta x}, \qquad (16.14)$$

$$\frac{\partial^2 u(x,t)}{\partial x^2} = \frac{u_{i+1}(t) - 2u_i(t) + u_{i-1}(t)}{\Delta x^2}. \qquad (16.15)$$

Using Eqs. (16.11) and (16.13), the spatially discretized system of Eqs. (16.1)–(16.3) reads

$$\frac{\partial u_i}{\partial t} = \frac{\tilde{D}_{i+1} - \tilde{D}_{i-1}}{2\Delta x}\frac{u_{i+1} - u_{i-1}}{2\Delta x} + \tilde{D}(x_i)\frac{u_{i+1} - 2u_i + u_{i-1}}{\Delta x^2}$$
$$- J_{fi}(u_i; v_i) - J_{so}(u_i) - J_{si}(u_i; w_i) + J_{\text{stim}}(x_i, t), \qquad (16.16)$$

$$\frac{\partial v_i}{\partial t} = \Theta(u_c - u_i)\frac{1 - v_i}{\tau_v^-(u_i)} - \Theta(u_i - u_c)\frac{v_i}{\tau_v^+}, \qquad (16.17)$$

$$\frac{\partial w_i}{\partial t} = \Theta(u_c - u_i)\frac{1 - w_i}{\tau_w^-} - \Theta(u_i - u_c)\frac{w_i}{\tau_w^+}, \qquad (16.18)$$

16.3 Numerical Calculation of the Action Potential

where it is implied that the discretized variables u_i, v_i, and w_i depend on time t. Note that we have added an extra term in Eq. (16.16) (the last term on the right-hand side) that depends both on time and space and represents the (normalized) stimulus current. The stimulus current is assumed to arise from physiological mechanism of the heart, and it is necessary for the excitation of the action potential. Here, the stimulus current $J_{\text{stim}}(x, t)$ is taken to be a periodic sequence of rectangular pulses of amplitude J_{amp} and duration τ_p. The period or basic cycle length between any two consecutive pulses T_p is taken to have values in the interval 800–1000 ms. The stimulus is further assumed to excite a small region around the left end of the cable (i.e., around $x = 0$) of length L_{exc}. For the temporal integration of Eqs. (16.16)–(16.18), we use a standard Runge-Kutta fourth-order algorithm with fixed time step Δt.

From Eqs. (16.16)–(16.18), we have calculated the ventricular action potential that is excited by a single rectangular pulse of the stimulus current. The total length of the ventricular tissue is $L = 3.5$ cm. The tissue is excited through its left end (i.e., at $x = 0$) using rectangular current pulses of different durations τ_p, and amplitude $J_{\text{amp}} = 4$ mA. The length of the excited tissue is $L_{\text{exc}} = 0.14$ cm. The results are shown in Fig. 16.3, monitored at two different locations on the cable, i.e., at $x \simeq 0.44$ cm (close to the excited region) and $x \simeq 1.75$ cm (in the middle of the cable). As it can be observed from the figure, the amplitude of the action potential as well as its width increases with increasing τ_p (from top to bottom). For $\tau_p = 6$ ms, the width of the action potential is about 160 ms that is quite close to reality. Left and right panels differ in that the latter exhibit a sharp peak at a time instant corresponding to the end of the stimulus current pulse. This sharp peak decreases until it practically vanishes for location on the cable relatively far from $x = 0$.

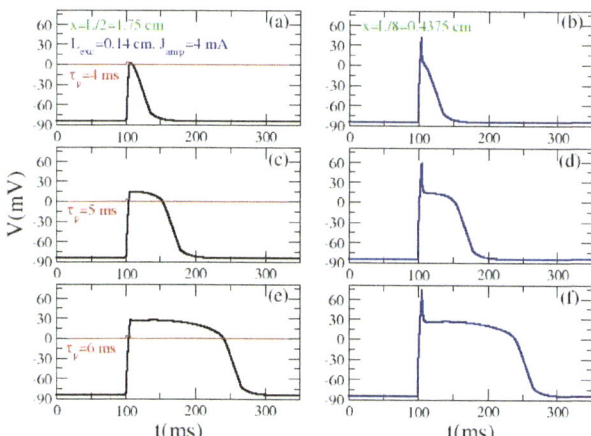

Fig. 16.3 Ventricular action potential excited by a single rectangular current pulse for $J_{\text{amp}} = 4$ mA, $L_{\text{exc}} = 0.14$ cm, and (**a**, **b**): $\tau_p = 4$ ms; (**c**, **d**): $\tau_p = 5$ ms; (**e**, **f**): $\tau_p = 5$ ms. The action potential is monitored at $x \simeq 0.44$ cm (right panels), and at $x \simeq 1.75$ cm (left panels)

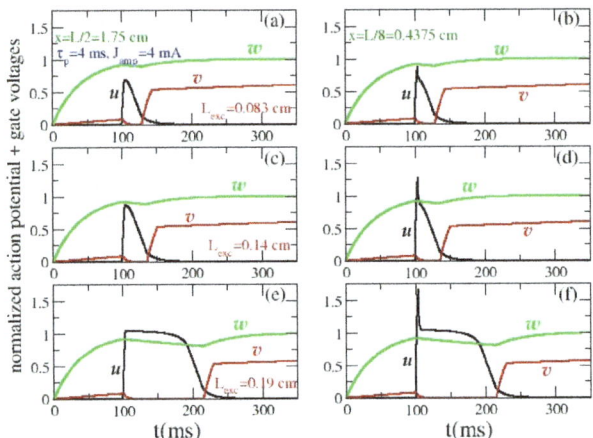

Fig. 16.4 Ventricular action potential u (black curves) and gating variables v and w (red and green curves, respectively) excited by a single rectangular current pulse with $J_{amp} = 4$ mA, $\tau_p = 4$ ms, and (**a**, **b**): $L_{exc} = 0.083$ cm; (**c**, **d**): $L_{exc} = 0.14$ cm; (**e**, **f**): $L_{exc} = 0.19$ cm. The variables u, v, and w are monitored at $x \simeq 0.44$ cm (right panels) and at $x \simeq 1.75$ cm (left panels)

Similarly, in Fig. 16.4, the normalized action potential u is monitored in time along with the gating variables v and w for $J_{amp} = 4$ mA, $\tau_p = 4$ ms, and three different lengths of the stimulated region of the cable, L_{exc}. Again, it is observed that the width (temporal duration) of the action potential increases with increasing L_{exc}. As in Fig. 16.3, the action potential is again monitored at two different locations on the cable, i.e., at $x \simeq 0.44$ cm and $x \simeq 1.75$ cm (left and right panels, respectively). In both Figs. 16.3 and 16.4, the action potential exhibits the right characteristics in (e) and (f) panels, as long as the shape and the width are concerned.

16.4 Numerical Evaluation of the Pseudo-Electrocardiogram

When one goes to the cardiologist, the first thing he/she does is an electrocardiogram (ECG). This is a relatively simple examination that measures the electrical activity of the heart and provides direct information that helps considerably in diagnosis and possible treatment of heart problems. Over the years, medical doctors have developed very strong, empirical rules that provide generally precise information on the various heart conditions. Arrhythmias are readily visible in the ECG as well as the presence of ischemia, infarction, etc. The ECG collects the electrical signals that propagate in the heart and delivers time dependent yet "averaged" information on these processes as seen from the exterior part of the heart.

The potential generated by the membrane voltage distribution within the virtual heart tissue is estimated using the expression [8–12]

16.4 Numerical Evaluation of the Pseudo-Electrocardiogram

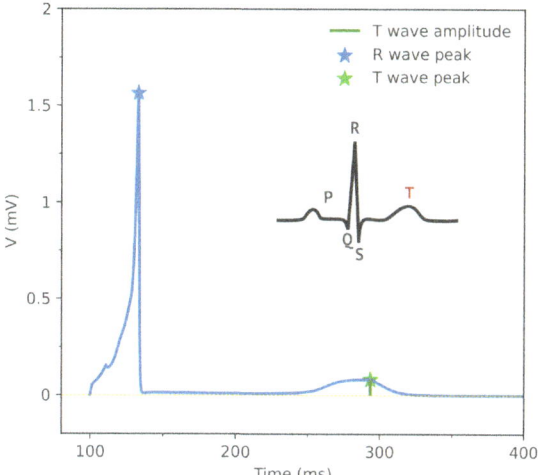

Fig. 16.5 Simulated pseudo-ECG (in the literature appears also as EKG) as a function of time t calculated using the three-variable Fenton-Karma model. For comparison, a drawing of a real ECG is shown in the inset. We can clearly detect the R and T wave equivalents whose amplitudes we designate with the blue and green stars, respectively

$$\Phi_e(\mathbf{x}^\star, t) = -K \int \nabla V(\mathbf{x}, t) \cdot \nabla \frac{1}{|\mathbf{x}^\star - \mathbf{x}|} d\mathbf{x}, \qquad (16.19)$$

where $\nabla V(\mathbf{x}, t)$ is the spatial gradient of the ventricular action potential, $K = 1.89\,\text{mm}^2$ is a constant number that depends on various electrophysiological quantities, and $|\mathbf{x}^\star - \mathbf{x}|$ is the distance from the a source point $\mathbf{x}$ to a field point $\mathbf{x}^\star$. This formula accumulates and propagates the electric dipole field in the heart model to the observation point. The time profile of Φ_e constitutes an approximation for the ventricular component of the ECG, i.e., the pseudo-ECG generated at a hypothetic electrode that was located at a particular distance away from the last epicardial cell along the cable. The translation from the mathematical pulse propagation to the effective ECG it generates may be seen in Fig. 16.5. We note that two prominent peaks appear that correspond to the R and T waves in the ECG. Other features are generally absent; this is generally due to the approximations made in the evaluation of the action potential propagation equation. In Fig. 16.6 we see in more detail how the ventricular ECG contributes significantly to the formation of the QRS cluster, as well as the T wave. For that reason the pseudo-ECG is expected to reproduce, at least qualitatively these features of an electrocardiogram.

For the results presented in Figs. 16.5, 16.6, we used the one dimensional version of Eq. (16.19) that reads

$$\Phi_e(x^\star, t) = -K \int \frac{\partial V(x, t)}{\partial x} \left(\frac{\partial}{\partial x} \frac{1}{|x^\star - x|} \right) dx. \qquad (16.20)$$

Then, for $x^\star = L$, we use the spatial profiles calculated from Eqs. (16.16)–(16.18) at each time instant, and we calculate $\Phi_e(L, t)$ that is the pseudo-ECG that was sought. The results of this procedure are shown in Fig. 16.6 along with the action potentials monitored at three different locations on the cable, for two set parameters. The first

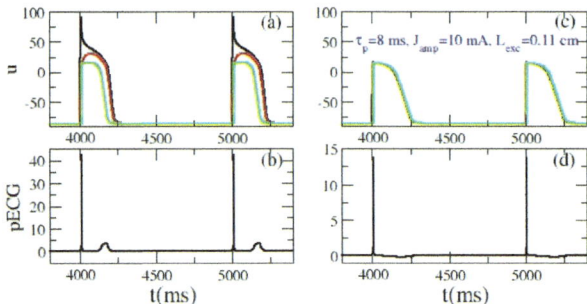

Fig. 16.6 Normalized ventricular action potentials u and pseudo-ECGs as a function of time t at three different locations on the cable, i.e., at $x = L/16$ (black curves), $x = L/4$ (red curves), and $x = L$ (green curves), for $J_{amp} = 10$ mA, $\tau_p = 8$ ms, $T_p = 1000$ ms, $L_{exc} = 0.16$ cm. (**a**, **b**) using the parameters of the Beeler-Reuter (BR) model (Table 16.1), and (**c**, **d**) using the parameters of the modified Luo-Rudy I (MLR-I) model (Table 16.1)

set corresponds to the Beeler-Reuter (BR) model shown in Fig. 16.6a and b, while the modified Luo-Rudy I (MLR-I) model in Fig. 16.6c and d. The action potentials shown in the upper panels are monitored at $x = L/16$ (black curves), $x = L/4$ (red curves), and $x = L$ (green curves). It is observed that the BR model provides action potential waves that vary with their location on the cable. That makes possible to obtain the corresponding pseudo-ECG (Fig. 16.6b) that exhibits clearly the R peak as well as the peak of the T wave. There are also indications of the S dip, which however cannot be observed in this scale. The R and T peaks are separated by a time interval of approximately 180 ms for the set of parameters of the BR model and the pacing parameters. On the other hand, the action potentials for the MLR-I model do not exhibit any significant dependence on their location on the cable. Consequently, the corresponding pseudo-ECG although it exhibits an R peak (but of much lower amplitude as compared to that from the BR model) does not exhibit a positive T wave peak. Instead, there is a shallow dip there, which, when it appears in a real ECG it indicates ischemic conditions [13].

16.5 Conclusions

Modeling the electrical activity of the heart through the use of precise mathematical methods enables the quantification of macroscopic and mesoscopic processes that take place in this valuable organ. This in turn leads to better understanding of the problems and ultimately helps in the cure of possible problems. The simple three-variable model introduced by Fenton and Karma applies primarily in the ventricle and is easily solvable numerically using different values for the parameters it includes [6]. Following the pulse propagation even in the simple one dimensional "cable" model is quite useful; from it we can determine a pseudo-ECG. The deterministic connection of the pulse transfer to the ECG is extremely important. One can connect problems in the propagation, stemming from heart tissue problems, to the electrocardiogram. Thus one can make a causal link between the propagation and the clinical observations. A good example is the case where as the pulse propagates it encounters an ischemic region, i.e., a sharp defect-like region where the conductivity is different from the average value. Numerical work in this case shows that for a certain range of values of the ischemic defect an inversion of the T wave in the ECG can take place. This is a definite cause and effect prediction that links the "microscopic" heart dynamics to the "macroscopic" ECG observation [13].

How can machine learning enhance the link between pulse propagation and electrocardiogram? One idea would be first to calculate many cases of heart propagation for various environments and determine the corresponding pseudo-ECGs. Subsequently, one could train a neural network with this information and then try to use it with observed ECGs. This approach would enable the determination of problematic regions in the heart that, in the case of 3D modeling, can give relatively precise spatial heart information.

16.6 Summary

- The Hodgkin-Huxley model can be specialized in the electrical pulse dynamics in the heart.
- Solution of the Fenton-Karma model gives quantitative information on the action potential propagation.
- A pseudo-ECG can be determined from the solution of the dynamical model that externally observed features of the propagation internally in the heart.

References

1. A.L. Hodgkin, A.F. Huxley, A quantitative description of membrane current and its application to conduction and excitation in nerve. J. Physiol. **117**(4), 500–44 (1952)
2. D. Noble, The Music of Life (Oxford University Press, Oxford, 2008)
3. A. Karma, Physics of cardiac arrhythmogenesis. Annu. Rev. Condens. Matter Phys. **4**, 313–37 (2013)
4. S. Alonso, M. Bär, B. Echebarria, Nonlinear physics of electrical wave propagation in the heart: a review. Rep. Prog. Phys. **79**, 096601 (56pp) (2016)
5. P.C. Franzone, L.F. Pavarino, S. Scacchi, *Mathematical Cardiac Electrophysiology* (Springer International Publishing, Switzerland, 2014)
6. F. Fenton, A. Karma, Vortex dynamics in three-dimensional continuous myocardium with fiber rotation: filament instability and fibrillation. Chaos **8**, 20 (1998); Erratum: Chaos **8**, 879 (1998)
7. J. Malmivuo, R. Plonsey, *Bioelectromagnetism: Principles and Applications of Bioelectric and Biomagnetic Fields* (Oxford University Press, Oxford, 1995)

8. K. Gima, Y. Rudy, Ionic current basis of electrocardiographic waveforms - a model study. Circ. Res. **90**, 889–896 (2002)
9. R.H. Clayton, A.V. Holden, Propagation of normal beats and re-entry in a computational model of ventricular cardiac tissue with regional differences in action potential shape and duration. Progress Biophys. Molecular Biol. **85**, 473–499 (2004)
10. O.V. Aslanidi, R.H. Clayton, J.L. Lambert, A.V. Holden, Dynamical and cellular electrophysiological mechanisms of ECG changes during ischaemia. J. Theoret. Biol. **237** 369–381 (2005)
11. K.Q. Wang, Y.F. Yuan, Y.Y. Tang, H. Zhang, Simulated ECG waveforms in long QT syndrome based on a model of human ventricular tissue. Comput. Cardiol. **33**, 673–676 (2006)
12. A. Bueno-Orovio, E.M. Cherry, F.H. Fenton, Minimal model for human ventricular action potentials in tissue. J. Theoret. Biol. **253**, 544–560 (2008)
13. E. Angelaki, N. Lazarides, G.D. Barmparis, I. Kourakis, M.E. Marketou, G.P. Tsironis, T-wave inversion through inhomogeneous voltage diffusion within the FK3V cardiac model. Chaos **34**, 043140 (2024)

Chapter 17
Machine Learning Cardiology

The Electrocardiogram Knows When You Have High Blood Pressure

Abstract Hypertension is a major medical problem that may lead to cardiovascular diseases. It is usually assessed through a periodic but systematic measurement of systolic and diastolic blood pressure. Since the electrocardiogram (ECG) is one of the most widely used diagnostic tools, it could be used for the initial evaluation of a patient suspected to have hypertension, if this information is actually contained in the ECG. We show that through the use of machine learning that this is indeed the case, i.e., that use of ECG may be used in detecting hypertension in a population without cardiovascular disease. The machine learning methods used involve logistic regression, k-nearest neighbors, random forest, and gradient boosting. Through this analysis, we find the basic clinical and ECG features that combined may give information on the hypertensive state of individuals and thus assist in early diagnosis and treatment.

17.1 Introduction

The application of machine learning algorithms in the management of data is transforming the landscape of different scientific fields, including clinical medicine. The fast-growing number of applications of machine learning and data analysis in healthcare allows identification of disease even in early stages and prognostication of clinical outcome, thereby increasing the efficacy of treatment options [1]. Artificial intelligence techniques have the potential to radically change the way cardiovascular medicine is being practiced, providing new tools to interpret data and make clinical decisions [2, 3]. AI offers opportunities not only to physicians to make more accurate and prompt diagnoses, but it can also identify hidden opportunities to improve patient management and avoid unnecessary spending. The electrocardiogram is one of the most widely used diagnostic tools; it is of paramount importance in the initial evaluation of a patient suspected to have a cardiovascular pathology. In this chapter we present a very practical application of machine learning in cardiology, viz. that of the detection of high blood pressure. While this can be done easily and routinely through a simple manometer, we follow a different approach. We may detect whether a person is hypertensive or not using features from their ECG in addition to anthropometric features such as age, sex, and body fat. This alternative approach, when successful, provides an alternative method to the age-old one. Additionally, with modern wearable technology, it could be implemented easily and cheaply.

17.2 Basics of the Electrocardiogram

The ECG measures voltage in time; the voltage is generated from the propagating pulse in the heart and since it is thankfully repetitive, so is the ECG trace. The typical ECG has twelve leads each measuring voltage in different body directions. In Fig. 17.1 we portray the time trace of a single lead focusing on the main pulse while also showing the one that follows after approximately 0.8 secs. The main waveforms of the trace are P, Q, R, S, T, and U, while the main time segments are also shown. The quantitative information presented in the time pulse of the ECG contains essentially all information necessary for understanding the health of the heart. The P wave starts in the upper part in the sino-atrial node where the electrical pulse initiates through cell membrane depolarization. The QRS complex is the central feature of the ECG and contains electrical propagation information in the ventricle. The T wave is a repolarization wave, while U wave is not always present. The time intervals between the waves, such as PR, QT, etc., depend on the speed of the pulse propagation, i.e., on the conductivity of the local biological matter as well as its geometry. Thus, an experienced cardiologist can deduce very valuable information about the physiological and dynamical states of the heart by scrutinizing the information contained in the ECG.

The electrocardiogram paper is a graph paper where for every five small (1 mm) squares you can find a heavier line forming a larger 5 mm square. The vertical axis

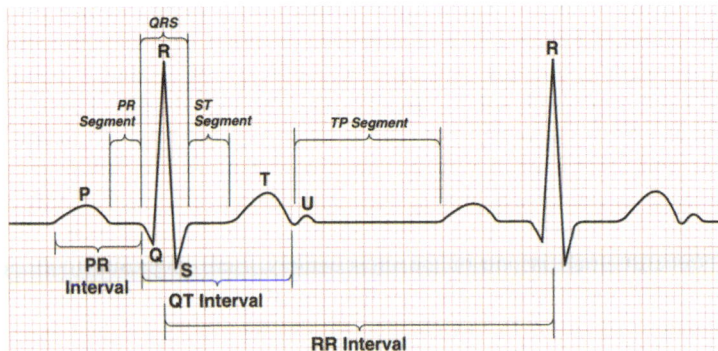

Fig. 17.1 Basic components of the ECG signal, including the P, QRS, ST, T, and U waveforms, the RR, PR, QRS, and QT intervals, and the PR, ST, TP segments. Image from [4] with permission

17.2 Basics of the Electrocardiogram

measures the heart's electrical pulse voltage that is measured in millivolts (mV). By standard, 10 mm in height equals 1 mV. Therefore each 1 mm square on the vertical axis equals 0.1 mV and each large square, 0.5 mV. The horizontal axis measures time. On a standard ECG the paper speed is 25 mm/s. Therefore, each 1 mm square on the horizontal axis equals 0.04 s, and each large square,0.20 s. Of the twelve leads six are limb leads (I, II, III, aVL, aVF, aVR) and six are precordial (V1, .., V6); each set focuses on different body planes [4].

The machine learning method we present is based primarily on random forests. This approach is very useful when we have binary type of selection processes. An individual is healthy or sick; we can portray this feature with a tree that has two branches. The "healthy branch" terminates since there is nothing more to consider there. In the other branch, we can start various questions and make the answers additional branches. This can go to various levels until it terminates. In Fig. 17.2 we show a general tree from an application in cardiology [5]. The problem here deals with a population that is tested for *left ventricular hypertrophy* of LVH. The tree uses binary but quantitative criteria for training and subsequently may be used in the form of a random forest with other potential patients for assessment.

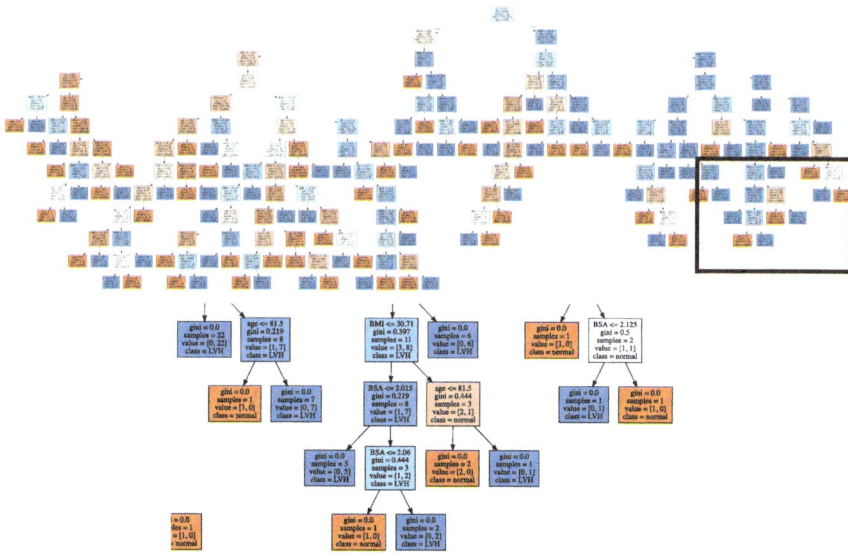

Fig. 17.2 (**a**) Depiction of a single decision tree out of many in a random forest. (**b**) The area inside the black box enlarged for visibility. The end nodes are leaf nodes, and their color denotes the class they represent. The specific tree is from a cardiology application that aims at detecting left ventricular hypertrophy (LVH): orange for LVH and blue for non-LVH [5]

17.3 Procedures and Data Handling

Traditionally physiological data are used by experienced medical doctors who with long years of experience know how to detect the hidden structures in the data and make judgments. Using machine learning instead or in addition to this classical method is a very new trend that still needs to show its full utility. If enough digital data are available and proper training takes place, then machine learning can definitely assist the medical professional and show hidden connections and help make an assessment. Sometimes it can also provide surprising results as is hypertension.

17.3.1 Clinical Procedure and Machine Learning

We focus on a particular study of hypertension detection in order to see how machine learning may be applied and provide useful information. In this study 988 subjects were enrolled of age larger than 17 years that did not show any essential hypertension and no other indications of cardiovascular disease. The details of the study can be found in ref. [6]. The clinical parameters used were the age, sex, the body mass index (BMI) as well as the percentage of body fat. The ECG data were taken with a digital 12-lead electrocardiogram machine in resting position with 10 s duration.

We tried three machine learning models that use Logistic Regression (LR), K-Nearest Neighbors (kNN) Classifier, and Random Forest (RF), using subsets of a number of clinical and electrocardiogram features [6]. In the former we have age, sex, body fat, etc., while in the latter group we used quantitative information obtained from the cardiogram. This information may be obtained from the individual cardiograms for each patient through an evaluation of the ECG quantities, such as all wave features, time lags between waves as well as empirically introduced combinations of these features. We mention the following specialized features: BMI-adjusted Cornell criterion, R wave amplitude in aVL, Cornell criteria, Area under R wave in lead I, QRS axis front, Corrected QT interval, P wave duration, PQ interval duration, QT interval duration, R wave amplitude in lead III, Planar frontal QRS-T angle, Area under R wave in aVF, Area under T wave divided by QRS complex area, Area under R wave in lead III, BMI-modified Sokolow-Lyon voltage, BMI-adjusted Sokolow-Lyon voltage, Total QRS area in all leads, S wave amplitude in V5, T wave amplitude in V5, S wave amplitude in V3, QRS complex duration, P wave amplitude in II, Area under QRS interval in V5, Q vs. S vector, J point deflection, Q wave duration, P axis in frontal plane, Intrinsicoid deflection in II, Area under S wave in V1, T wave duration, T wave amplitude in III. This long list of features is selected using long experience of cardiologists and also experimental "trial-and-error" efforts with the data.

17.3 Procedures and Data Handling

In order to obtain the specific numbers for the features for each individual, we use the digitized cardiogram and extract the basic quantities such as heights in millivolts, time duration in milli-seconds, areas under the curves in $mV \cdot ms$, slopes in degrees, etc., from the representative beat it generates. After the initial quantification of the signal, we formed standard combinations of these figures that are used in cardiology. These include the Cornell criterion, corrected QT interval, BMI-adjusted Sokolow-Lyon voltage, etc. All these were used as features for the cardiogram of each individual.

In the random forest method, we form an ensemble of decision trees. Each decision tree performs a series of binary decisions (splits) by selecting a subgroup of the input features (such as age, body fat, BMI), effectively trying out different feature order and feature combinations. A random forest builds a large collection of de-correlated trees and then averages their votes for the predicted class. They are good predictors even with smaller datasets due to a technique called bootstrap aggregating (bagging). Bagging trains multiple trees on overlapping, randomly selected subset of the data, and makes the final decision based on the votes of the different trees. For modeling random forests we used the Python package scikit-learn. We optimized the model parameters by minimizing the random forest's built-in out-of-bag error estimate that is almost identical to that obtained by N-fold cross-validation. This technique enables random forests to be trained and cross-validated in one pass.

17.3.2 Feature Engineering and Feature Selection

The Python code may process the additional ECG waveform measurements from the 1 *sec* representative beat produced by the electrocardiograph. Using the automated measurements provided by the machine as a starting point, we calculated several other features such as areas, slopes, and heights under curves. Some of the features that proved most important for our final model are: (a) The BMI-adjusted Cornell criteria: product of R in aVL + S in V3 and BMI. (b) The BMI-modified Sokolow-Lyon (SL) criteria: BMI (in continuous value) divided by the sum of the amplitudes of S wave on V1 and R wave on V5. It has been shown that body mass affects the amplitude of the R and S waves, as the electrical currents cover different distances. (c) Amplitude of R wave in aVL lead.

Random forests are capable of handling nonlinear interactions as well as correlations among features. Initially we had 60 features in our dataset, but through feature selection one can reduce their number and retain the very essential ones. We perform feature selection since after initial testing we find that the model performed slightly better when trained with a reduced set of features. Additionally, some of the features exhibited high correlation among them, as assessed by Pearson's correlation test, and thus keeping only one of them retains the essential information while providing a clearer picture of the feature's contribution [6] (Fig. 17.3).

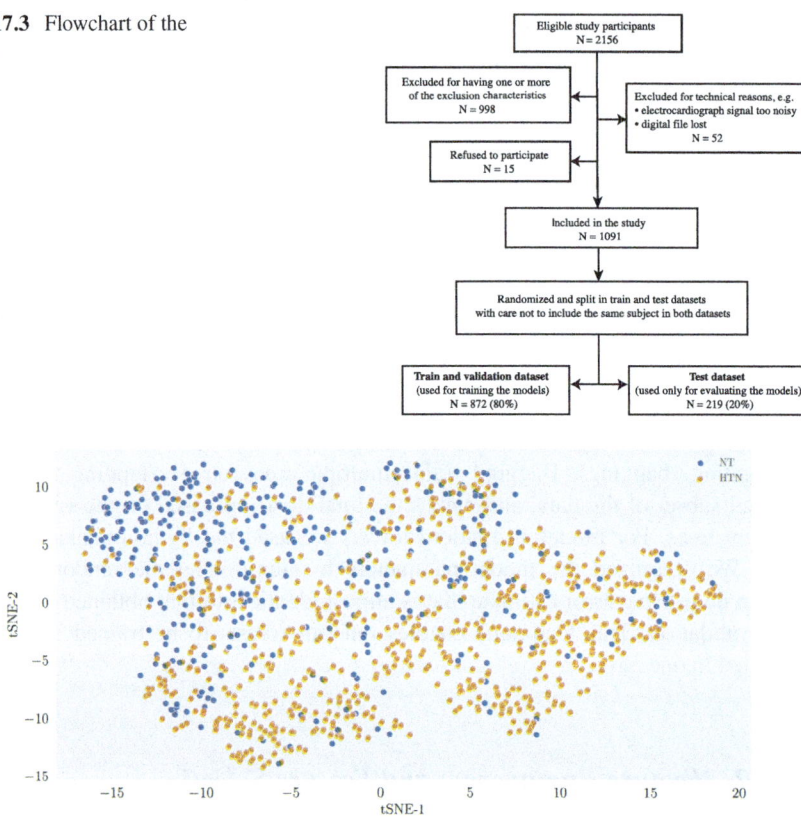

Fig. 17.3 Flowchart of the study

Fig. 17.4 Study subject clustering using t-distributed Stochastic Neighbor Embedding (t-SNE). NT signifies the normotensive participants and HTN the hypertensive. The axes of the two dimensional space are given in arbitrary units

17.3.3 Datasets and Feature Importance

The dataset was split into a train set (80%), used directly to learn the parameters of the model, and a test set (20%) that was used exclusively for final performance evaluation of the models. For validation we used the random forest's internal out-of-bag (oob) set. Stratification for sex and hypertension, while splitting, ensured balanced sets with same proportions of these features in train and test sets as in the original dataset. All reported performance results are on the test set. Feature importance graphs are also on the test set, as using the train set inflates the importance of some features that might not be as important in predicting the outcome.

We visualized subsets of the anthropometric and ECG features using $t-SNE$. As shown in Fig. 17.4, each point is a participant characterized by the following set of features: age, body fat, BMI-adjusted Cornell criteria, R wave amplitude in aVL, and BMI-modified Sokolow-Lyon voltage (BMI divided by SV1 and RV5). This particular subset of features seems to visually separate hypertensive patients, who are represented by dots mostly on the upper left corner, from normotensive patients, which are the ones in the rest of the plot. On the basis of this as well as quantification through a rank correlation test, we conclude that the best subset to train our final models was the one consisting of age, sex, body fat, BMI-adjusted Cornell criteria, R wave amplitude in aVL, and BMI-modified Sokolow-Lyon criteria.

17.4 Results

Explaining predictions from tree models is always desired and is particularly important in medical applications, where the patterns uncovered by a model are often more important than the model's prediction performance. Scikit-learn's tree ensemble implementation allows for the computing of measures of feature importance. These measures aim at providing insight into which features drive the model's prediction. We calculate the feature importance metric called SHAP (SHapley Additive exPlanations), an approach used to explain the output of any machine learning model. The SHAP metric connects optimal credit allocation with local explanations using the classic Shapley values from game theory and their related extensions. Visualizing feature importance using SHAP values is thought to be more accurate for global and local feature importance, i.e., when importance calculated on each feature instead of all of them.

The random forest model's accuracy on detecting hypertension was 84.2% with specificity 66.7% and sensitivity was 91.4%, while the area under the receiver operating characteristic curve (AUC/ROC) was 0.86, for the standard decision threshold of 0.5. By moving the threshold to 0.6, we may increase the specificity to 78.0% without sacrificing the sensitivity too much (new value was 84.0% versus 91.4%). The results for all our models are summarized in Table 17.1. Feature importance calculated by SHAP is shown in Fig. 17.5a. Dependence on BMI-adjusted Cornell criteria is shown in Fig. 17.5b. The horizontal dashed line represents the cut-off between having a negative effect on being hypertensive (below the line) and a positive one (above the line). On the x-axis, we see that participants with a value approximately above $37\,\mathrm{mV} \cdot \mathrm{kgr/m^2}$ have a positive chance of being hypertensive. These values were calculated by SHAP on the random forest model.

The outcome of the machine learning analysis and the results presented in Table 17.1 is that indeed with proper blend of clinical and electrocardiogram features it is possible to detect hypertension with reasonably high probability. This is done without the use of the standard pressure gauge machine. The clinical features are clearly very important; however, they alone cannot give the full picture. The information from the electrocardiogram participates very strongly in the final assessment of hypertension. This clearly means that in some way the "mechanical" heart pressure information is contained in the propagation features of the electrical pulse in the heart.

Table 17.1 Performance metrics for random forest, logistic regression, and k-nearest neighbors. For each method, we used different training features

Model	Features	Accuracy (%)	AUC(ROC)	Sensitivity (%)	Specificity (%)
Random forest 1 (thresh.= 0.6)	Age, sex, BF, BMI-adj-Cornell, R in aVL, BMI-modified SL	**84.2**	**0.89**	**84.0**	**78.0**
Random forest 1	Age, sex, BF, BMI-adj Cornell, R in aVL, BMI-modified SL	84.2	0.86	91.4	66.7
Random forest 2	Age, sex, BF, BMI-adj-Cornell, R in aVL	82.0	0.86	90.0	66.0
Random forest 3	Age, sex, BF, BMI-adj Cornell, R in aVL, BMI-modified SL, BMI	83.2	0.867	91.4	63.2
Logistic regression	Age, sex, BF, BMI-adj Cornell, R in aVL, BMI-modified SL	77.8	0.77	93.6	39.7
K-nearest neighbors	Age, sex, BF, BMI-adj-Cornell, R in aVL, BMI-modified SL	78.8	0.87	97.6	40.2

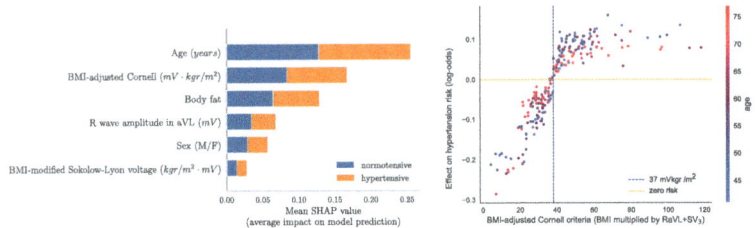

Fig. 17.5 Results in detecting hypertension by the random forest. (**a**) Feature importance calculated on the test set using Shapley Additive explanations (SHAP). (**b**) Effect of BMI-adjusted Cornell criteria on the risk of being hypertensive. Each dot in the plot is a participant whose BMI-adjusted Cornell value is indicated on the x-axis. The values on the y-axis effectively indicate the effect of each participant's set of features in characterizing them as hypertensive

17.5 Conclusions

In this chapter we applied basic machine learning techniques in cardiology. Even though this topic might be surprising in the context of nonlinear physics and mathematics, we know already from the previous chapter that processes in the heart are truly complex. On the other hand, the data collection is done through electrocardiography that provides a cheap but potent tool that has been used for over hundred years in medicine. The medical doctors have thus accumulated very long experience on its use and the connection to various problems in the heart. The question AI poses is whether its application in cardiology may bring more information to doctors so that they can perform more detailed and accurate diagnosis. Furthermore, there is also the possibility that information is somehow convoluted in the ECG and not readily available to human eye; here AI can help and bring it into the foreground.

In the work we presented we showed that proper use of machine learning in ECG can uncover a somehow surprising feature, viz. that the electrocardiogram contains at least in a partial form information on hypertension. The latter is usually assessed through "mechanical" techniques, such as the use of spigmomanometer, and it is not associated with the electrocardiogram. We saw, however, that traces of this information are embedded in the electrical heart information and with proper care we can extract it. In some sense this is not unreasonable since hypertension is related to blood pumping by the heart and clearly the specifics of this appear to be reflected in the heart's electrical currents. The detection of hypertension from clinical and ECG data can be though as a "triumph" of machine learning. Since it is able to connect seemingly unrelated information and produce knowledge that is not routinely connected with this type of data, this success opens up another possibility that is even more interesting. One may think to use this simple ECG technique, collect electrical heart data, and attempt to link the directly to specific heart problems in precise heart locations. In order to accomplish this feat, however, one has to use precise computational models for the heart that are based on physical and possibly chemical modeling. This *physics informed machine learning* that couples physical modeling with data may open up a new possibility in assessing the microscopic workings of the heart and other medically interesting setups [7].

17.6 Summary

- Machine learning can extract seemingly hidden information from the electrocardiogram.
- The electrocardiogram contains information on hypertension.
- Physics informed machine learning could produce knowledge of the heart conditions from simple external measurements.

References

1. A. Haque, A. Milstein, L. Fei-Fei, Illuminating the dark spaces of healthcare with ambient intelligence. Nature (London) **585**(7824), 193–202 (2020)
2. K.C. Siontis, X. Yao, J.P. Pirruccello, A.A. Philippakis, P.A. Noseworthy, How will machine learning inform the clinical care of atrial fibrillation. Circ. Res. **127**(1), 155–169 (2020)
3. K. Seetharam, S. Raina, P.P. Sengupta, The role of artificial intelligence in echocardiography. Current Cardiol. Rep. **22**(9), 50 (2020)
4. A. Goldberger, Z.D. Goldberger, A. Shvilkin, *Goldberger's Clinical Electrocardiography*, 9th edn. (Elsevier, Amsterdam, 2018)
5. E. Angelaki, M.E. Marketou, G.D. Barmparis, A. Patrianakos, P.E. Vardas, F. Parthenakis, G.P. Tsironis, Detection of abnormal left ventricular geometry in patients without cardiovascular disease through machine learning: an ECG-based approach. J. Clin. Hypertens. **00**, 1–11 (2021)
6. E. Angelaki, G. D. Barmparis, G. Kochiadakis, S. Maragkoudakis, E. Savva, E. Kampanieris, S. Kassotakis, P. Kalomoirakis, P. Vardas, G.P. Tsironis, M.E. Marketou, Artificial intelligence-based opportunistic screening for the detection of arterial hypertension through ECG signals. J. Hypert. **40**, 2494 (2022)
7. M. Raissi, P. Perdikaris, G.E. Karniadakis, Physics- informed neural networks: a deep learning framework for solving forward and inverse problems involving nonlinear partial differential equations. J. Comput. Phys. **378**, 686–707 (2019)

Chapter 18
Epidemiology with Physics Informed Machine Learning

COVID-19 Predictions with Machine Learning

Abstract The COVID-19 pandemic took the whole earth by surprise in early January 2020 and soon affected almost all countries in the world. The spreading of the virus became a critical factor in the implementation of various measures, at times draconian, in order to curtail it. Multiple efforts focused on predictions of the infection rates using available data. We present here two data-driven efforts that used a physics motivated approach and led to a successful predictive power.

18.1 Introduction

The SARS-CoV-2 or COVID-19 coronavirus stared in the Wuhan province of China and soon spread all over the world. The first wave that arrived from China in the early time of 2020 caused a relatively small fatality rate but with a strong increasing tendency. The public and the authorities were not prepared for such an epidemic and did not know initially what precautions to take against the virus. During this early period the virus was very strong and potent due to the lack of immunity in the public. One critical and urgent issue for science was how to use existing data together with general epidemiological models in order to predict the evolution of the virus spreading and more importantly its termination. In China the virus appeared on December 23, 2019 in the Wuhan region, and due to its fast spreading strict rules of social distancing were imposed almost a month later. Most of the rest of the world adopted similar harsh social distancing measures but with a time delay, starting in February or March or even later in 2020 depending on the country. There was a substantial delay of the order of 2–3 months for the imposition of measures outside of China. During this period we have access to quantitative information on the evolution of the virus in China in a period that includes the onset of social distancing measures.

If we mentally bring ourselves to March 2020 outside of China and wish to use data tools in order to predict the evolution of the pandemic in our own countries, we have at our disposal two pieces of information: The first is the hard infection data of China during the first 3 months of the virus spreading that include in the early part no measures and in the later part the strict social distancing measures. If these data are the footprint of the pandemic dynamics in China we might reasonably expect that the evolution will be similar in other countries provided we take into account the specifics of each country. The second, independent, piece of information comes from the theory of epidemics, the "physics" or "mathematics" of the viral propagation. The epidemic may be modeled by a simple SIR model that describes very succinctly the actual dynamics through susceptible (S), infected (I), and removed (R) populations [1]:

$$\frac{dS}{dt} = -\alpha S I \tag{18.1}$$

$$\frac{dI}{dt} = \alpha S I - \mu I, \tag{18.2}$$

where $S \equiv S(t)$, $I \equiv I(t)$, and R are the percentage of susceptible, infected, and removed (deceased or recovered) individuals, respectively, and the infection and removal rates $\alpha \equiv \alpha(t)$ and $\mu = \mu(t)$, respectively, are functions of time in general. We note that $S + I + R = 1$, and thus the equation for $R(t)$ is not really necessary. For the present initial analysis that aims in the prediction of the actual evolution of the epidemic in each country, we will make the reasonable assumption of a constant removal rate $\mu = \mu_0$, while the time dependence of the infection rate will play a crucial role in the data-assisted predictions. This is because the presence of social distancing restrictions affects directly the infection rate. Since the epidemic phenomenon was delayed in most other countries, we may use a China data-physics approach, viz. utilized the data from China, and through the nonlinear set of the SIR equations we make predictions for other countries assuming that the process of spreading is "universal" and independent of local particularities [2].

18.1.1 China Data and Early COVID-19 Predictions

In Fig. 18.1 we show the infected population data that were available during the end of March 2020 [3]. We notice that the data increase to a maximum and then decrease smoothly. There is an outlier for one specific date where the reported data are more than four times larger. The Chinese authorities discarded this measurement at that time, and thus we will also ignore it in the present analysis. What is truly remarkable is that the data can be fitted reasonably well by a Gaussian function in time. Usually we fit a variety of spatial or other non-time dependent data by a Gaussian function, but fitting the phenomenon with a time-Gaussian process is novel. Since this function fits the China epidemic data in this period, we may safely assume that a similar general dependence will follow in other countries as well. The functional form is

$$N_I(t) = A e^{-\frac{(t-\mu_I)^2}{2\sigma_I^2}},$$

18.1 Introduction

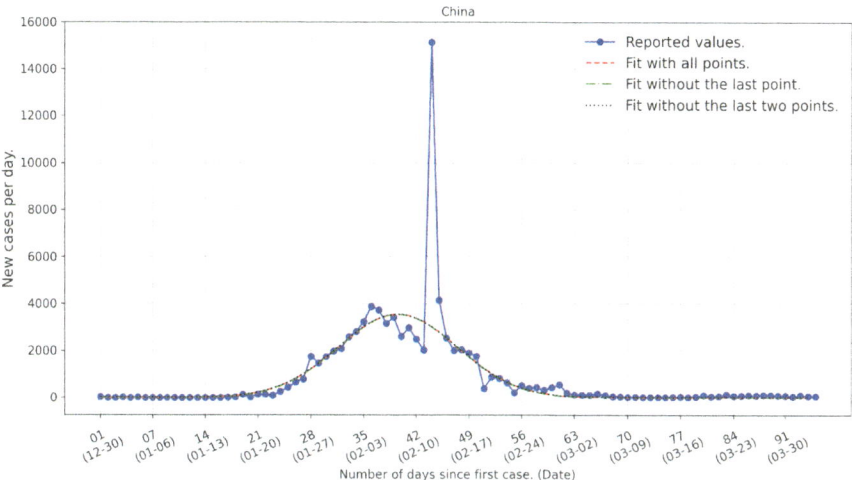

Fig. 18.1 The number of individuals reported to have been infected with COVID-19 in China during the period December 31, 2019 to March 31, 2020. We notice a large outlier on February 13, 2020 related to reporting issues; this singular event has been excluded from the analysis. We find a good fit with a time-Gaussian function with mean 40.5 days, standard deviation 7.9 days, and height equal to 3557 cases. From this fit we determine an infection horizon at 4σ approximately equal to 2 months from the start of the infection

where $N_I(x)$ is the number of new infected persons each day, while t is the time in days since the first event in the country we consider, China or other. The three fitting parameters A, μ_t, and σ_t determine the height, the position of the peak, and the width of the Gaussian, respectively. The China data give $A = 3557$ cases, while $\mu_t = 40.5$ and $\sigma_t = 7.9$ days.

In order to make then predictions for other countries, we assume a similar Gaussian-in-time distribution for the COVID pandemic, use the available early data, and fit it to the Gaussian. In the fitting procedure we initialize each fitting parameter with a randomly assigned value within a reasonable range of values. We then utilize simulated annealing to determine the global minimum of the mean absolute error (MAE) between the reported values (RV) and the predicted ones (N_I) each at day j:

$$\text{MAE} = \frac{1}{N} \sum_{j=1}^{N} |RV(j) - N_I(j)|, \qquad (18.3)$$

where N is the total number of days from the first reported case until the latest day of the available data. The process of minimization is performed iteratively—when MAE reduces compared to a previous step the parameters A, μ_t and σ_t are registered and kept. Subsequently, Gaussian random numbers with the new parameters are used as simulated infected data and are compared to the available infected country data, and the process of MAE stochastic minimization is repeated. The iteration

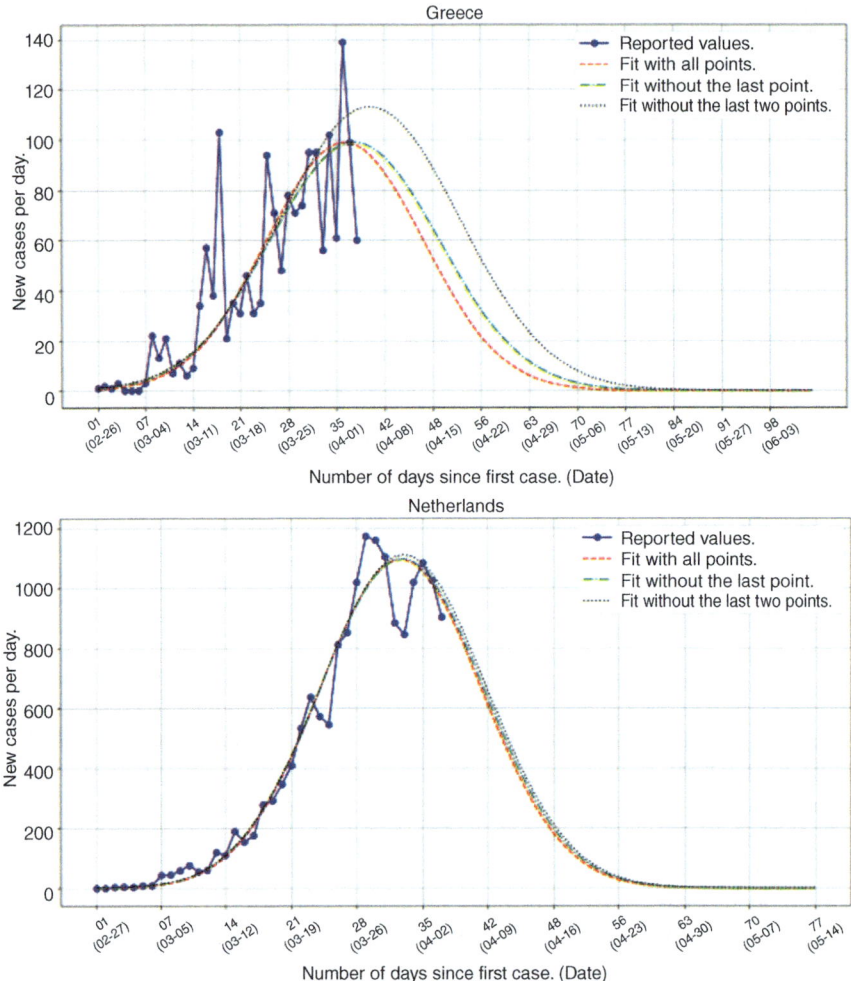

Fig. 18.2 Greece and Netherlands: infection data (blue) and time-Gaussian fits. This analysis gives a definitive horizon for the first wave of the pandemic

stops when the three parameters converge to the optimal ones. The outcome of this procedure produces a predictive landscape for the evolution of the infected persons in different countries. In Fig. 18.2 we show the evolution and prediction of two countries, Greece and Netherlands [3]. We observe that the Gaussian fit produces a clear prediction for the evolution of the infection many days after the time of observation while giving a clear indication of its termination.

18.1.2 Time-Gaussian Pandemic Evolution

The critical assumption for the projection of the evolution of the infected persons is that of the time-Gaussian evolution; the latter involves a well-rounded rise to a maximum and an almost symmetric decay in the number of infected persons. Although the China data were the guide for the Gaussian evolution, we should be able to recover it from the simple SIR model. If we look at the SIR equations, we notice from the second equation that the condition for flattening the infection growth, i.e., $\dot{S} = 0$, occurs at the critical susceptible number $S_c = \mu/\alpha$. For $S > S_c$ the infected population will grow, while at $S = S_c$ it will reach a maximum, and then as $S(t)$ drops further it will decay to zero. In this case the infection ends with the individuals fully recovered or removed from the population. The values thus of the two parameters α and μ are critical for the precise mode of the evolution of the infection.

The value of α determines how aggressive is the infection; for large values a large population is infected, and only at a small number of susceptibles the infection decays. The value of μ on the other hand controls the rate at which the individuals are removed from the infection process; large values of μ result in a very fast decay of the infection. A typical evolution is shown in Fig. 18.3 on the left where the infected population is seen to grow fast and reach a maximum and subsequently have a relatively slow decay. The time evolution is distinctly non-Gaussian. In this figure we assumed a constant infection rate α while $\mu = \mu_0$ a constant as well.

We now assume that specific measures are taken in the processes of the infection; this can be easily implemented in the SIR model by taking the infection rate to be explicitly time dependent with a specific functional decay. The effect of the implementation of social distancing leads unavoidably to a decay in the contacts and

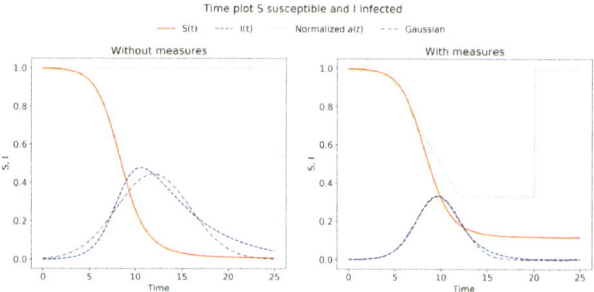

Fig. 18.3 Time evolution of susceptible S(t) (red solid line), infected I(t) (blue dashed line) fractions, and the normalized infection rate (green dotted line) together with the corresponding Gaussian approximation for the I(t) (black dashed dotted line). (**a**) Constant infection rate with no social distancing measures. The Gaussian parameters are height = 0.445, mean = 11.92, standard deviation = 3.94, and PCC = 0.964, and (**b**) time-varying infection rate that introduced social distancing measures. The Gaussian parameters are height = 0.330, mean = 9.70, standard deviation = 2.44, and PCC = 0.998

thus the infections. Thus we may take for the time dependence of $\alpha(t)$ an initially constant rate that gives rise to infection growth, then a precipitous linear decline attributed to social distancing measures, then a saturation at a new lower level, and finally the return to "freedom," i.e., to the value before the imposition of measures. The $\alpha(t)$ time dependence as well as the results of this SIR process, are seen in the right panel of Fig. 18.3. We notice the presence of a Gaussian-like evolution for the infected population.

We see that the actual time evolution of the infected population is not only distinctly Gaussian but, more importantly, there is not even a damped recurrence of the infection as well. It is noteworthy that this rather optimistic scenario of measure imposition that gradually decreases social distancing and thus the infection rate is what approximately happened in most countries. We quantify the correlation between the infection curve and the fitted Gaussian function using the Pearson Correlation Coefficient (PCC) that measures the statistical relationship between two curves (PCC = -1 means perfectly anticorrelated curves, PCC = 0 non-correlated and PCC = 1 perfectly correlated curves). In the case without measures the PCC between the curve of the infection and the fitted Gaussian equals 0.964 while with measures is equal to 0.998 indicating an almost perfect match between the two curves.

A simple mathematical indication for the Gaussian-in-time-like in time behavior of the infection evolution can be obtained from an approximate analysis of Eqs. (18.1) and (18.2). We note that the imposition of gradual social distancing through a linear drop in the infection rate leads, to the lowest order, to a linear drop in the susceptible population. Assuming thus a dependence of the form $S(t) = \gamma - \beta t$, where β, γ are appropriate parameters that are constants to the lowest order, we obtain a solution for the infected population that is practically Gaussian, i.e.,

$$I(t) = I_c e^{-\frac{\alpha\beta}{2}t^2 + (\alpha\gamma - \mu)t}, \qquad (18.4)$$

where I_c is the initial infection rate at time negatively large for the present form of the solution. We can argue that the additional exponential dependence in the approximate solution of Eq. (18.4) is very weak leading to the Gaussian exponential form found also through the numerical analysis and shown in Fig. 18.1.

The analysis of data shows empirically that the China virus infection, at least in the first phase, followed a Gaussian-in-time evolution. Although this feature appeared initially to be at odds with the simple SIR model, it nevertheless follows from it when gradual social distancing measures are imposed. Furthermore, provided that the gradual measure imposition is retained, the mathematical model does not predict an infection recurrence. The latter is however possible when the social distancing measures are lifted. This simple Gaussian model can be used in order to predict the evolution of the infection in other countries where the pandemic was a bit delayed compared to China. The Gaussian nature of the evolution is directly attributed to the time dependence of the infection rate induced by social distancing.

18.2 COVID-19 Predictions with Physics Informed Machine Learning

In the previous section we found that the infection rate $\alpha(t)$ in the SIR model is the quantity that controls the dynamics of the infection. The presence or even the modification of social distancing measures has a direct bearing in the infection and the population dynamics. However this function is not known a priori, and thus the usefulness of the SIR model in predicting is limited. We may however "turn the problem around," and by using existing infection data we may extract $\alpha(t)$ and thus predict more accurately. This is a physics-based neural network approach that was applied to the second wave of the infection [4, 5]. Specifically, we train the network using every time the new available data, for instance, on a weekly basis. The combined use of the SIR equations with this data is used in order to extract the specific time dependent infection rate for the time period at hand. Using this infection rate, the dynamical equations, and the last available data as initial conditions, we can make short or even longer term predictions. The former can be of the order of a week while the latter longer. The accuracy of the predictions depends critically on the *continuity* of the measures. Sudden and abrupt changes need some time to be learned.

In order to facilitate the analysis through the combination of physics with data, we transform the equations of the SIR model into a single equation in a form that is more workable. The aim of the math is to arrive at an equation for the time dependent infection rate that can be learned by the network. In order to facilitate and accelerate the data learning process, we can use pretrained data generated through the SIR model for various parameters. Further simplification is the linear assumption in the decay of the infection rate; we thus need to learn essentially a single parameter that is the slope—designated as σ—of $\alpha(t)$. Once the infection rate is known, we use country specific data, validate the resulting SIR model, and vary σ to observe the changes in the epidemic.

Although more complex functional forms of infection rate can be taken, the effective linearization of $\alpha(t)$ with slope σ simplifies the computational process while capturing the essential phenomenon and gives a simple quantitative estimate of the imposed measures. The values of the slope σ are obtained directly through machine learning and thus are completely data-driven. Each infection curve is associated with an effective decay slope σ that gives the overall efficiency that the measures exercise on the spreading. Once the PINN is trained with a certain phase data, it can be used to predict the evolution of the infection at a later time period.

18.2.1 Mathematical Manipulations

In order to blend physics with data we need to use a model that describes reasonably well the dynamics of the problem at hand. Many times simplicity is a virtue, and thus the very basic SIR model is a good starting point. It not only incorporates the basic actors in the epidemic process but also is simple and mathematically and computationally easily tractable. Additionally, as we saw in the previous section, it does have good predictive power. We start with Eqs. (18.1) and (18.2) and assume that the infection and removal rates $\alpha(t)$ and $\mu(t)$, respectively, are general functions of time. In order to make the mathematics more transparent, we introduce a new variable $q(t)$ as follows:

$$I(t) = e^{q(t) - \int_0^t \mu(t')dt'}. \tag{18.5}$$

Upon substitution to the set of Eqs. (18.1 and 18.2), we obtain

$$\dot{S} = -\alpha e^{q-\nu} S \tag{18.6}$$

$$\dot{q} = \alpha S \tag{18.7}$$

$$\nu \equiv \nu(t) = \int_0^t \mu(t')dt'. \tag{18.8}$$

Using Eqs. (18.6 and 18.7) we obtain a closed equation for q, i.e.,

$$\ddot{q} = -\left(\alpha e^{q-\nu} - \frac{\dot{\alpha}}{\alpha}\right)\dot{q}. \tag{18.9}$$

The single, highly nonlinear, second order equation, Eq. (18.9), describes fully the dynamics of the SIR infection model for arbitrary time dependence of both the infection and removal rates. In the case of constant infection and removal rates, it can be solved exactly [2].

Since we need to capture the infection rate dynamics in terms of the infection data, we need to have a differential equation for $\alpha(t)$. Starting with the general equation (18.9) and assuming for simplicity constant recovery rate $\mu(t) \equiv \mu$, we proceed in the simplification of Eq. (18.9) to $\nu = \mu t$ and thus the infected rate to $I(t) = \exp[q(t) - \mu t]$. It is simple to derive the first order equation for the infection rate starting from Eq. (18.9) [5]:

$$\frac{d\alpha}{dt} + f(t)\alpha = g(t)\alpha^2 \tag{18.10}$$

$$f(t) = -\frac{\ddot{q}}{\dot{q}} \tag{18.11}$$

$$g(t) = e^{q - \mu t}. \tag{18.12}$$

Equation (18.10) is useful since we are interested in the inverse problem of finding the infection rate from the data. The specific form of $\alpha(t)$ determines the infection evolution. We know, from the previous section, that a monotonic linear

18.2 COVID-19 Predictions with Physics Informed Machine Learning

drop in the infection rate, as for instance introduced by gradual social distancing measures, leads to an approximately Gaussian evolution.

For the analysis of the first wave we considered the case where the infected population behaves as a Gaussian function in time. Let us follow quantitatively the consequences of this assumption keeping in the Gaussian exponent not only a quadratic but also a linear time-term; the latter introduces some time asymmetry. Assuming then the exponent in $I(t)$ to be

$$q(t) = \beta t^2 + \gamma t, \qquad (18.13)$$

we obtain after some algebra the following analytical expression [5]:

$$\alpha(t) = (2\beta t + \gamma)\left[\alpha(0)\gamma + \frac{2\beta}{2\beta t + \gamma} - (\gamma - \mu)e^{\beta t^2 + (\gamma - \mu)t}\right]. \qquad (18.14)$$

In Eq. (18.14) the dominant term is that of the linear decay since at long times, and for $\beta < 0$ the Gaussian term in Eq. (18.14) practically disappears, while the exponential term also decays when $\mu > \gamma$. This analytical result shows how closely connected is the shape of the infection time function to the functional form of the infection rate.

18.2.2 Machine Learning Procedure

After this preliminary analysis we bring neural networks in the foreground. For each country to be studied we make a model that uses a deep neural network with five layers, each having 100 nodes, all with a sigmoid activation function. There is a unique output node. Since we have the SIR equations at our disposal, we use them in order to pretrain the model; the simulated SIR data are produced with an arbitrary linear function for $\alpha(t)$ and a constant value μ. For the training we use a loop that minimizes the mean squared error loss on the data, MSE_D:

$$MSE_D = \frac{1}{N_D} \sum_{i=1}^{N_D} |x_i - \tilde{x}_i|^2, \qquad (18.15)$$

where $\{x_i, \tilde{x}_i\}_{i=1}^{N_D}$ denote the set of the reported and predicted cases, with $x_i = ln(I_i)$ the former and $\tilde{x}_i = model(t_i)$ the corresponding predicted ones. The mean squared error loss is defined through Eq. (18.9) with an explicit functional form for the infection rate:

$$\alpha = \alpha(t) = \sigma_0 + \sigma t. \qquad (18.16)$$

This form together with a constant removal rate μ leads to $\nu = \mu t$ and $\tilde{x} = q(t) - \mu t$ and finally to the second explicit mean square error MSE_{SIR}:

$$\text{MSE}_{SIR} = \frac{1}{N_{SIR}} \sum_{j=1}^{N_{SIR}} |f(t_j, \tilde{x}_j, \dot{\tilde{x}}_j, \ddot{\tilde{x}}_j, \sigma_0, \sigma, \mu)|^2, \quad (18.17)$$

where

$$f(t_j, \tilde{x}_j, \dot{\tilde{x}}_j, \ddot{\tilde{x}}_j, \sigma_0, \sigma, \mu) = \ddot{\tilde{x}}_j + \left(\alpha_j e^{\tilde{x}_j} - \frac{\dot{\alpha}_j}{\alpha_j}\right)(\dot{\tilde{x}}_j + \mu) = 0. \quad (18.18)$$

We notice that the mean square error tries to minimize through the deep neural network model the proximity of the data to the optimal SIR equation that describes parametrically the data. This is the true blending of physics with data in the context of PINNs.

In practical terms we use the following algorithm: Step I: For each country under consideration we load the real (weakly) data and smooth it using a seven time step moving average. Subsequently we scale it using Min-Max normalization. Step II: We load each pretrained country model and tune all weights by minimizing both the MSE_D and MSE_{SIR} loss functions. Step III: In the end of this process determine the optimal values for $\alpha(t)$ and μ for the given country. We use the pretrained model in order to accelerate the training process for each country and early stopping with 100 epoch horizon. The machine learning algorithms are implemented in Python using TensorFlow/Keras and the ADAM optimizer. The data used for the study are published online by *OurWorldInData.org*.[1] In Fig. 18.4 we show a graphical summary of the flowchart for the implementation of the procedure.

After extracting the optimal $\alpha(t)$ and μ for each country, we use them in order to solve the SIR model, Eqs. (18.1 and 18.2). The solution is then fitted to the country's real data using the initial conditions (I_0, S_0) as fitting parameters. The total number of the predicted cases during the "first wave" period of each country, including the relative error to the corresponding total number of reported cases and the total number of cases obtained by varying $\alpha(t)$ by $\pm 10\%$, is presented in Table 18.1. A plot of the results for each country is shown in Fig. 18.5. In the map of Fig. 18.6 we portray the results of Table 18.1 in a more graphical way.

18.2.3 Short-Term Predictions

We used the arsenal of physics with AI in order to extract dynamical parameters such as the time dependent infection and the removal rates from the documented

---

[1] https://ourworldindata.org/

18.2 COVID-19 Predictions with Physics Informed Machine Learning

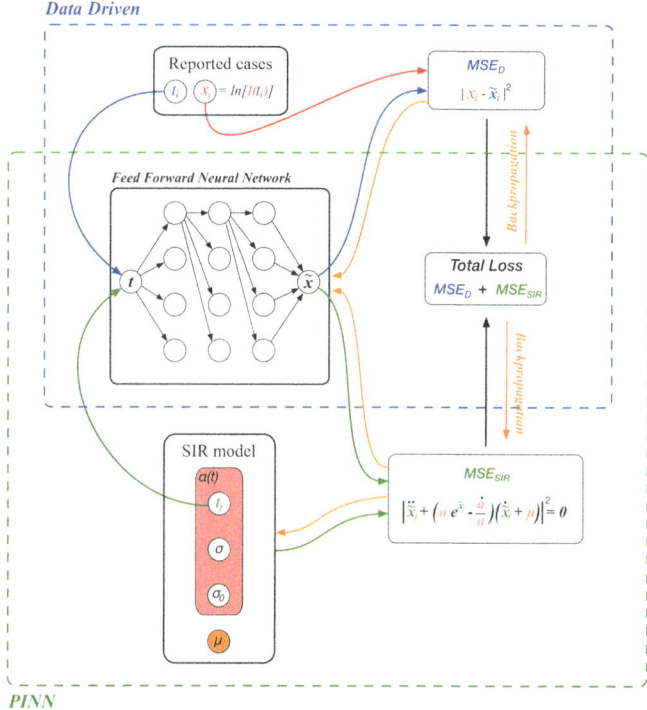

Fig. 18.4 A graphical summary of the physics informed method used for training the model

infection data of the first infection wave. This led to the quantitative comparison of the efficiency of measures in different countries in the first battle against the COVID menace. Through SIR pretraining and real data training, we obtain a good handle on the infection information. The next challenge is to turn this knowledge into a predictive mechanism for short and intermediate range. One idea is the following: Since we obtain the value of $\alpha(t)$ for a given range until a certain day, we may assume that the last values will persist for a short future horizon, for instance, that of a week. We may then implement this information in the SIR model; with the additional use of the last infection data as initial conditions, we can predict the evolution for the future. The past data then produce the parameters and the physics model the short-term future predictions. This process can be naturally repeated for instance on a weekly basis, and thus have a regular prediction for the evolution of the pandemic. We show the results in Fig. 18.7 with data for both the present phase of the pandemic and the *prediction* obtained for a horizon of one week. For the predictions we used in the PINN training the data after the end of the first pandemic wave. We also show the comparison with the real post-prediction data; we notice that the network's short-term predictive power is quite good on average in most countries.

Table 18.1 Left: the total number of reported cases during the "first wave" for each country and the corresponding predicted cases and percentage error obtained from our model, including the predictions with ±10% variation of a(t). Right: a bar plot of the slope $-\sigma$ of each country signifying the degree of adherence to measures. The higher the bar, the more reduced is the transmission rate due to the control measures

Country	Total cases Reported	Total cases Predicted	Error (%)	$\alpha(t) + 10\%$ (% Difference)	$\alpha(t) - 10\%$ (% Difference)
USA	1961185	1945830	-0.8	1793214 (-7.8)	2063029 (6.0)
Italy	240961	275667	14.4	259349 (-5.9)	284978 (3.4)
Spain	245938	280859	14.2	249833 (-11.1)	298190 (6.2)
UK	286141	312211	9.1	217528 (-30.3)	382591 (22.5)
Germany	186839	215563	9.5	182532 (-15.3)	233021 (8.1)
The Netherlands	50412	55040	9.2	52524 (-4.6)	56583 (2.8)
France	149668	163580	9.3	117539 (-28.2)	187154 (14.4)
Greece	2967	3014	1.6	2794 (-7.3)	3155 (4.7)

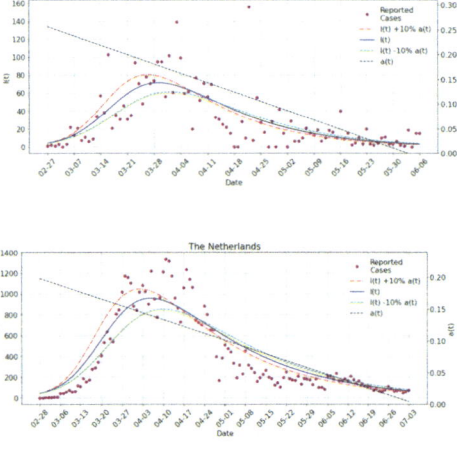

Fig. 18.5 Prediction on infection rate $I(t)$ through the extracted $\alpha(t)$ for Greece and the Netherlands. Magenta dots represent the reported cases of each country. Red dashed and dotted line, blue solid, and green dashed line represent the infections with +10%, no change, -10% to the infection rate $\alpha(t)$, respectively. Black dashed line denotes the extracted infection rate for the two countries

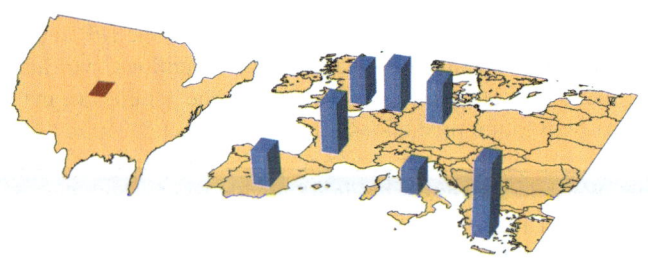

Fig. 18.6 Presentation of the values -σ derived for the first phase of the infection for each country on a map for better visualization. The red color in the USA denotes opposite sign, i.e., in terms of the present interpretation nonadherence to measures on average. In the European countries the value of Greece is maximal

18.2 COVID-19 Predictions with Physics Informed Machine Learning

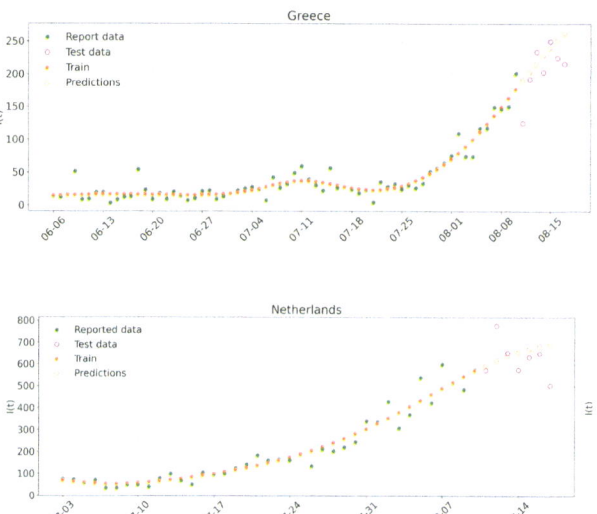

Fig. 18.7 Short-term predictions using PINN's and comparison with existing data. We use the second wave COVID-19 data up to a date to train the network, make the prediction for the following week from the last date, and subsequently compare with the infections occurred. The percentage error in the total number of predicted cases is 4.5% for NL and 10.5% for GR

18.2.4 The Case of Greece

We applied the PINN method to the weekly data from Greece during the period August 18, 2020 until December 18, 2020, i.e., for a period of four months. We made predictions every Friday evening with a horizon of one week. In Greece, there was a lockdown imposed on November 7, 2020. In Fig. 18.8 we present the comparison of the weekly PINN model predictions to the actual reported data after each prediction period, including the weekly total reported and predicted cases and the relative error. Additionally, we show longer prediction horizons based on data in different periods of the pandemic evolution. We observe that the model is quite adoptive to the data behavior, and it gives a quite good short-term average prediction, when the infection rate does not change rapidly. It also demonstrates the degree of effectiveness of the measures. On the contrary, the model does not follow the pandemic's evolution when the infection rate changes very rapidly, either increasing, as during the week between October 19 and October 26, or decreasing, i.e., immediately after the lockdown on November 07. The large increase in the prediction error is due to an abrupt imposition of measures exactly that week. This action causes a very sudden change in the infection rate that cannot be accounted for immediately from the model. In the following weeks the model adopts, and the short-term predictive power increases again.

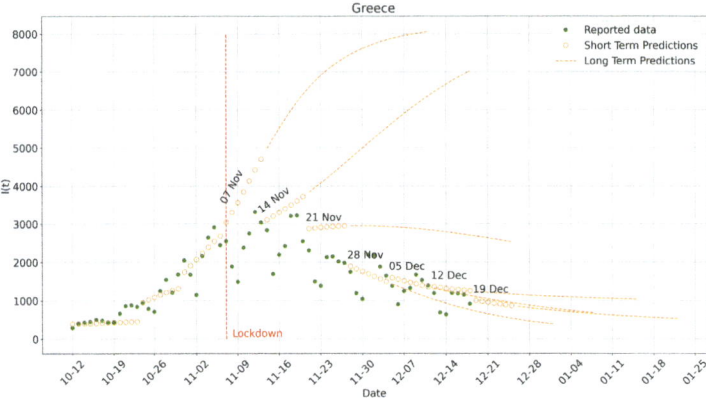

Fig. 18.8 The PINN machine learning model applied to Greece using infection data (green filled circles) for a period of approximately 4 months. The predictions are weekly (open orange circles) or longer term (dashed lines). The trend of the infection is generally captured by the predictions. Sudden lockdowns change $\alpha(t)$ very rapidly, and the short-term prediction becomes less accurate. The intermediate term predictions demonstrate clearly the effectiveness of the lockdown while predicting the observed slow decay of the virus spreading. The dates on the graph give the date the intermediate prediction was made

18.3 Conclusion

Epidemics do not respect boarders or races and involve a very large number of individuals. The COVID-19 waves appeared very suddenly and caused tremendous hardship and death to numerous persons in almost every part of the Earth. As in most cases in life, knowledge is the real defensive power. Thus very quickly in the pandemic it became essential to be able to predict the evolution of the virus spreading. This helped the policymakers to take appropriate decisions that helped curtail the spreading. From the point of view of science we are facing here a complex dynamical problem that depends on two aspects, data and models. Using infection data alone is a pure empirical approach that is clearly very limited by the specifics and the errors in the data. On the other hand, using theory without data may provide some general understanding of the processes involved, but this approach cannot have any realistic predictive power. Thus, one needs to use both data and theory in a way that hand in hand can provide the framework for both general understanding and quantitative predictions. A data-driven but physics inspired approach is a framework that combines both.

During the first phase of the spreading different countries took different measures and the epidemic evolved differently. In most cases where social distancing was imposed, the Gaussian law found in the China data was also followed. The time scale and the specifics of the infection dynamics were of course different in each country. It is thus important to be able to assess the effectiveness of the measures imposed if possible with a single index parameter. To do this we viewed the available data under

18.3 Conclusion

the prism of the very simple SIR epidemic model with time dependent coefficients. We found that a linear drop in the infection rate leads to the time-Gaussian infection pattern.

In order to extract the time dependent infection rate from the data, we used physics informed neural networks, i.e., a machine learning method that uses input from the actual model assumed, viz. SIR. This input, together with the real infection data from each country we considered, led to a prediction of the assumed linear in time infection rate. The data derived slope σ signifies the adherence of each country to social distancing. In Greece, for instance, the slope is large in absolute value, designating strong application of the imposed measures by the individuals. In the other extreme, we find the USA with a practically zero slope, demonstrating that the measures taken for the first pandemic wave had low efficiency. The other six countries we analyzed fall in intermediate locations between these two extremes. Application to the SIR model of each country, an alternative infection rate that differs by a few percent ($\pm 10\%$) in total from the one obtained through machine learning, gives an estimate of how dependent the infection is on the applied measures. We find that this variation, while it affects the early SIR fast rise strongly, results in quite a different infection decay and horizon in countries like the UK.

Once we know how the PINN behaves with the data for the initial period of the infection, we can use it for the second phase that starts after the practical decay of the first wave. The use of the cumulative data for each country leads to an $\alpha(t)$ that is then used in the SIR model for predictions. After the end of the prediction horizon—typically one week—we compare the prediction data with the real data. We find that while the short-term predictive power of PINN is good, it has large deviations in countries where the data appear to have a rather stochastic character.

The basic conclusion is that the use of physics informed ML enabled the extraction of COVID-19 infection information in different countries and showed how different measures and practices are directly reflected in the data and ultimately make predictions. The use of physics in machine learning gives specificity to the data but, on the other hand, is restricted and sometimes limited to inserted physics knowledge. The present approach assumes a well-mixed, essentially uniform country, an assumption that is introduced through the use of the SIR model. However, countries have regions, and each region may behave differently for geographical, environmental, cultural, and population reasons. If regional data are available, one can go one step further and introduce spatial in addition to temporal distribution in the infection and from this be able to obtain more accurate results and predictions. We believe the methodology used in this work may be extended in this more realistic case and provide a more direct approach to local dynamics and the effectiveness of imposed measures at a local level.

The present approach depends on the SIR model and the fact that naturally we do not know the time dependent infection rate $\alpha(t)$ during the prediction period. As a result we assume that during the prediction period the infection rate is that of the previous appropriate time segment or at least it does not change rapidly. One could envision using other epidemiological models for the physics informed machine learning or even more sophisticated neural network architectures than the feed-forward networks, i.e., the recurrent neural networks, such as the Long Short-Term Memory (LSTM) model, and make other projections for the rate $\alpha(t)$. These could lead to improved prediction accuracy in similar cases of interest.

18.4 Summary

- Data together with simple assumptions lead to useful predictions in complex processes such as epidemics.
- The simple SIR model captures the essentials of epidemics.
- Physics informed machine learning applied in the COVID-19 epidemic has good predictive power.

References

1. W.O. Kermack, A.G. McKendrick, A contribution to the mathematical theory of epidemics. Proc. R. Soc. Lond. A Contain. Pap. Math. Phys. Char. **115**, 700 (1927)
2. L. Hufnagel, D. Brockmann, T. Geisel, Forecast and control of epidemics in a globalized world. Proc. Natl. Acad. Sci. USA **101**, 15124 (2004)
3. G.D. Barmparis, G.P. Tsironis, Estimating the infection horizon of COVID-19 in eight countries with a data-driven approach. Chaos Solit. Fract. **135**, 109842 (2020)
4. M. Raissi, P. Perdikaris, G.E. Karniadakis, Physics- informed neural networks: a deep learning framework for solving forward and inverse problems involving nonlinear partial differential equations. J. Comput. Phys. **378**, 686 (2019)
5. G.D. Barmparis, G.P. Tsironis, Physics-informed machine learning for the COVID-19 pandemic: adherence to social distancing and short-term predictions for eight countries. Quantit. Biol. **10**, 139 (2022)

Part VII
Conclusion

Chapter 19
Foundations

Statistical Mechanics of Learning

Abstract In the foundations of artificial intelligence we find many scientific disciplines, but it appears that statistical ideas from physics have played an important role. We give a brief summary of some of these ideas and concepts that have led the way to the modern developments. In particular we introduce and describe briefly the Ising model, a fundamental model in statistical mechanics that incorporates interactions among individual spins. We then use the Ising model to introduce spin glass models that include additionally disorder in the inter-spin interactions and lead to novel ideas on statistical phase transitions. These ideas take us to the Hopfield networks of neurons that model associative memories through the application of the Hebbian rule for learning. The Boltzmann machines are a further refinement of this type of complex nonlinear lattice models that have learning properties. We conclude with the protein folding problem that constitutes a dynamical search in a statistically immense space of possible states that can be actually implemented through artificial intelligence algorithms.

19.1 Introduction

On Tuesday, October 8, 2024, the Secretary General of the Royal Swedish Academy of Sciences announced the recipients of the Nobel Prize in Physics for 2024 and sent seismic waves in the global scientific community. The recipients were John Hopfield and Geoffrey Hinton, two well-known scientists, one a physicist by training while the second a computer scientist. If this was not enough, the following day, October 9, we found out that Demis Hassabis and John Jumper were two of the three recipients of the chemistry prize—they were not chemists but computer scientist as well. While the physics prize was awarded for the engagement in developing the area of artificial intelligence, the chemistry prize went to a very important application of AI in biology.

Awarding scientific work in the area of artificial intelligence should have been expected in retrospect given the explosive development and penetration of AI in all areas of science, engineering, and technology. Since the topic of this book is the connection of artificial intelligence with complexity, it is then pertinent to ask whether complex systems have played some role in the works that led to these exciting developments. The answer is affirmative, and in fact the Physics Nobel Prize of 2021 awarded in part to Giorgio Parisi honored research in this general area. The Nobel citation notes that "around 1980, Parisi discovered hidden patterns in disordered complex materials. His discoveries are among the most important contributions to the theory of complex systems. They make it possible to understand and describe many different and apparently entirely random materials and phenomena, not only in physics but also in other, very different scientific disciplines, such as mathematics, biology, neuroscience and machine learning" [1]. The work in *spin glasses*, i.e., complex disordered magnetic systems that have glassy properties, quoted by the Nobel committee have led to numerous theoretical and practical developments not only in physics but also in artificial intelligence. In this chapter we will perform a time reversal and go back in time in order to describe briefly the statistical foundations that led to these phenomenal developments of the year 2024.

19.2 The Ising Model

The Ising model is a fundamental mathematical model in statistical mechanics, used to describe ferromagnetism and phase transitions. It was originally proposed in 1920 by Wilhelm Lenz and later solved in one dimension by his student Ernst Ising in 1925 [2]. Lenz proposed the Ising model as a way to understand magnetic materials, where atoms in a lattice interact with their neighbors. Each atomic "spin" can take one of the two values, $+1$ (spin up) or -1 (spin down), representing the possible orientations of a magnetic moment [3]. Ernst Ising solved the model in one dimension in his doctoral thesis and found that there is no phase transition in a 1D chain of spins at any nonzero temperature. This result led Ising to incorrectly conclude that the model would not exhibit a phase transition in higher dimensions, limiting its initial impact on physics [4].

Interest in the Ising model was revived in 1944 when Lars Onsager provided an exact solution for the two dimensional (2D) model on a square lattice without an external magnetic field. Onsager's solution demonstrated a phase transition at a nonzero temperature, marking a critical development in the study of phase transitions [5]. In the decades following Onsager's solution, the Ising model became a key tool in statistical physics. It has been extended to study various phenomena, including the behavior of alloys, liquid-gas systems, and neural networks [6]. The model also laid the groundwork for the Renormalization Group (RG) approach developed by Kenneth Wilson, which revolutionized the understanding of critical phenomena [7]. The Ising model is widely used not only in physics but also in disciplines like computer science, neuroscience, and social science. It is a powerful example of a system that can exhibit complex collective behavior and phase transitions all arising from simple, local interactions.

Fig. 19.1 One dimensional Ising model

19.2.1 Mathematical Foundation of the Ising Model

The Ising model is a mathematical model used to describe magnetic interactions in a lattice of spins. Each spin can have one of the two states, represented as $+1$ (up) or -1 (down). The spins are arranged in a lattice, typically in one, two, or three dimensions. The model captures how neighboring spins interact to create macroscopic magnetization.

In the simplest one dimensional (1D) Ising model, we consider N spins arranged in a line. Each spin s_i interacts only with its nearest neighbors Fig. 19.1. The Hamiltonian H, representing the system's energy, is given by

$$H = -J \sum_{i=1}^{N-1} s_i s_{i+1} - h \sum_{i=1}^{N} s_i, \tag{19.1}$$

where J is the coupling constant. If $J > 0$, the interaction is ferromagnetic (favoring aligned spins), and if $J < 0$, it is antiferromagnetic (favoring anti-aligned spins), h is the external magnetic field acting on each spin, and $s_i = \pm 1$ is the spin state at site i.

In the two dimensional (2D) Ising model, the spins are arranged on a square lattice. Each spin interacts with its four nearest neighbors. The Hamiltonian becomes

$$H = -J \sum_{\langle i,j \rangle} s_i s_j - h \sum_i s_i, \tag{19.2}$$

where $\langle i, j \rangle$ denotes the summation over nearest neighbor pairs.

The partition function Z that is used in determining the thermodynamic properties of the system is given by

$$Z = \sum_{\{s\}} e^{-\beta H}, \tag{19.3}$$

where $\beta = \frac{1}{k_B T}$, with k_B being the Boltzmann constant and T the temperature, and $\{s\}$ denotes the set of all possible spin configurations.

The probability of a particular spin configuration $\{s\}$ is

$$P(\{s\}) = \frac{e^{-\beta H(\{s\})}}{Z}.$$

The magnetization M of the system, which measures the average spin alignment, is defined as follows:

$$M = \left\langle \sum_{i=1}^{N} s_i \right\rangle,$$

where $\langle \cdot \rangle$ denotes the thermal average. The susceptibility χ, indicating the system's response to an external magnetic field, is given by

$$\chi = \frac{\partial M}{\partial h}.$$

For the 2D Ising model, a phase transition occurs at a critical temperature T_c when $h = 0$. Below T_c, the system exhibits spontaneous magnetization even without an external field. For the 1D Ising model, no phase transition occurs at finite temperatures.

19.3 Spin Glasses

Spin glasses are disordered magnetic systems characterized by random interactions between spins, leading to complex energy landscapes and slow dynamics. The study of spin glasses began in the 1970s with work by S. F. Edwards and P. W. Anderson, who introduced the Edwards-Anderson (EA) model to describe these systems. This model incorporated randomness into the interactions between spins, allowing for both ferromagnetic and antiferromagnetic couplings [8]. The EA model became a foundational model for studying disorder in magnetic systems.

Shortly after the EA model, David Sherrington and Scott Kirkpatrick developed the mean-field version of the spin glass, now known as the *Sherrington-Kirkpatrick (SK) model*. The SK model introduced the concept of replica theory to handle the randomness in spin glass systems [9]. The replica method became a significant tool in statistical mechanics, though initially it led to inconsistent results due to assumptions of replica symmetry. A major breakthrough occurred in 1979 when Giorgio Parisi developed a solution to the SK model using *replica symmetry breaking (RSB)*. Parisi's solution introduced the idea of multiple replica states, capturing the complexity of the spin glass energy landscape [10].

The concepts developed from spin glass theory, such as frustration and rugged energy landscapes, have found applications in fields beyond physics. Spin glass models have been used to study neural networks, optimization problems, and complex systems in computer science [11].

19.3.1 Ising Spin Glass Systems

Ising spin glasses are disordered magnetic systems in which the interactions between Ising spins are random and can be either ferromagnetic favoring aligned spins or antiferromagnetic that favors anti-aligned spins. The Hamiltonian H of a spin glass model, such as the Edwards-Anderson model in d-dimensions, is given by

$$H = -\sum_{\langle i,j \rangle} J_{ij} s_i s_j - h \sum_i s_i, \tag{19.4}$$

where
$s_i = \pm 1$ represents the spin at site i, J_{ij} are random coupling constants that can take positive or negative values, often modeled as Gaussian random variables with mean $\langle J_{ij} \rangle = 0$ and variance $\langle J_{ij}^2 \rangle = J^2$, and h is an external magnetic field.

In spin glasses, conventional magnetization M is not suitable to describe the system's state due to the randomness of interactions. Instead, the order parameter q, called the overlap parameter, measures the similarity between two equilibrium states $\{s_i^{(1)}\}$ and $\{s_i^{(2)}\}$:

$$q = \frac{1}{N} \sum_{i=1}^{N} s_i^{(1)} s_i^{(2)}. \tag{19.5}$$

This parameter reflects the degree of similarity (or "overlap") between different configurations.

19.3.1.1 Replica Theory and Free Energy

The replica method is used to compute the average free energy of spin glasses, particularly in the mean-field Sherrington-Kirkpatrick model. By replicating the partition function Z n-times, we can express the free energy F as

$$F = -k_B T \lim_{n \to 0} \frac{\langle Z^n \rangle_J - 1}{n}, \tag{19.6}$$

where $\langle \cdot \rangle_J$ denotes averaging over the disorder, i.e., over the random couplings J_{ij}.

In the SK model, the *Parisi solution* introduced the concept of replica symmetry breaking (RSB), where the overlap parameter q takes on a continuous distribution. This RSB approach captures the complexity of the energy landscape in spin glasses and reflects the system's many metastable states. Spin glasses exhibit a phase transition at a critical temperature T_c, below which the system enters a spin glass phase with nonzero overlap q. Above T_c, the system behaves like a paramagnet with $q = 0$.

19.4 The Hopfield Model

The *Hopfield model* is a recurrent neural network model introduced by John J. Hopfield in 1982. It serves as a mathematical model of associative memory, enabling the storage and retrieval of patterns based on partial or noisy inputs. The Hopfield model is notable for its connections to both neuroscience and statistical mechanics, laying foundational work in the field of artificial neural networks. In his seminal paper, Hopfield presented a network of binary neurons, where each neuron is connected to every other neuron through symmetric connections. The system's energy landscape, inspired by the Ising model from statistical mechanics, allows it to settle into stable states, which serve as memory patterns. By minimizing an energy function, the model retrieves the closest stored pattern corresponding to a given input [12] (Fig. 19.2).

The state of the Hopfield network can be described by binary variables $s_i = \pm 1$ for $i = 1, \ldots, N$, where N is the number of neurons. The connection strength, or synaptic weight, between neurons i and j is denoted by w_{ij}, and the dynamics of the network are governed by an energy function:

$$E = -\frac{1}{2} \sum_{i \neq j} w_{ij} s_i s_j. \tag{19.7}$$

Hopfield showed that this energy function decreases with each update, guiding the network toward stable states (local minima), which correspond to stored memory patterns.

The Hopfield model has been extended in various ways to increase storage capacity and improve memory retrieval. Extensions like the Boltzmann machine, introduced by Geoffrey Hinton and Terrence Sejnowski, build on the probabilistic framework of the Hopfield model but incorporate stochastic dynamics [13]. The model has also inspired associative memory applications in image recognition and error correction [14]. The Hopfield model remains a fundamental model in theoretical neuroscience, providing insights into how biological systems store and retrieve memories. Modern neural network architectures, such as transformers and deep learning models, have roots in concepts established by the Hopfield model, particularly in areas of pattern recognition and attractor dynamics [15].

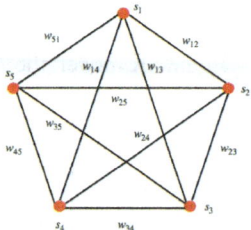

Fig. 19.2 A simple Hopfield network with five spin nodes denoted with bullets and with symmetric connections $w_{ij} = w_{ji}$

19.4.1 Mathematical Formulation of Hopfield Networks

A Hopfield network is a recurrent neural network model introduced by John Hopfield that stores information as attractors, allowing associative memory through stable states in an energy landscape. The Hopfield network consists of N neurons, each represented by a binary state s_i, where $s_i = \pm 1$. Each neuron i is connected to every other neuron j by a symmetric weight w_{ij}, so that

$$w_{ij} = w_{ji}.$$

The stability of the network is described by an energy function E, defined as

$$E = -\frac{1}{2}\sum_{i=1}^{N}\sum_{j=1}^{N} w_{ij} s_i s_j + \sum_{i=1}^{N} \theta_i s_i, \tag{19.8}$$

where w_{ij} is the weight between neuron i and j, s_i is the state of neuron i and θ_i is the threshold of neuron i. The energy function E drives the network dynamics, pushing it toward local minima, which represent stored patterns.

19.4.1.1 Weight Update Rule (Hebbian Learning)

To store a set of patterns $\{\xi^\mu\}$ (where $\mu = 1, 2, \ldots, P$), the weights w_{ij} are set using the *Hebbian learning rule*:

$$w_{ij} = \frac{1}{N}\sum_{\mu=1}^{P} \xi_i^\mu \xi_j^\mu, \tag{19.9}$$

where ξ_i^μ is the state of neuron i in pattern μ, and P is the number of patterns stored.

This rule allows the network to learn multiple patterns, with an approximate storage capacity of $0.15 \times N$.

19.4.1.2 State Update Rule

The state of each neuron i is updated asynchronously using

$$s_i(t+1) = \operatorname{sgn}\left(\sum_{j=1}^{N} w_{ij} s_j(t) - \theta_i\right), \tag{19.10}$$

where sgn is the sign function, outputting $+1$ if the argument is positive and -1 otherwise, $s_j(t)$ is the state of neuron j at time t, and θ_i is the threshold for neuron i.

The network updates each neuron until it reaches a stable state, corresponding to a local minimum in the energy landscape.

The Hopfield network converges to stable states, or attractors, which represent the stored patterns. It can store up to approximately $0.15 \times N$ patterns before retrieval errors occur, while the symmetric weights ($w_{ij} = w_{ji}$) ensure that the energy function decreases monotonically, leading to guaranteed convergence.

19.5 Boltzmann Machines

The *Boltzmann machines* are a class of stochastic recurrent neural networks named after Ludwig Boltzmann. They were introduced by Geoffrey Hinton and Terrence Sejnowski in the mid-1980s as a model for associative memory and learning. Boltzmann machines utilize a probabilistic approach to find optimal solutions in complex search spaces, making them useful for optimization and unsupervised learning tasks [16]. The model builds on the Hopfield network by introducing stochastic elements, allowing the system to escape local minima in its energy landscape through thermal noise. This stochasticity enables the Boltzmann machine to explore a broader range of configurations, leading to improved learning capabilities (Fig. 19.3).

The energy of a Boltzmann machine with N units, where each unit s_i is binary ($s_i = 0$ or 1), is defined by

$$E = -\sum_{i<j} w_{ij} s_i s_j - \sum_i b_i s_i, \tag{19.11}$$

where w_{ij} is the symmetric weight between units i and j, and b_i is the bias for unit i. The probability of a state $\{s\}$ is given by the Boltzmann distribution:

$$P(\{s\}) = \frac{e^{-\beta E(\{s\})}}{Z},$$

where β is the inverse temperature, and Z is the partition function, ensuring normalization.

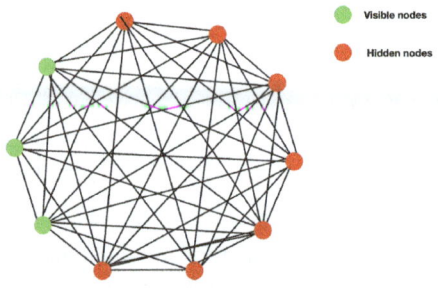

Fig. 19.3 A fully connected Boltzmann machine with visible and hidden nodes

19.5 Boltzmann Machines

Training a Boltzmann machine involves adjusting the weights w_{ij} and biases b_i to minimize the energy of states corresponding to the desired outputs. The training uses Gibbs sampling and *contrastive divergence* methods introduced by Hinton [17], which improve the efficiency of weight updates in the presence of stochastic units.

In 1986, Smolensky introduced the *restricted Boltzmann machine (RBM)*, a simplified version with a bipartite structure, where units are divided into visible and hidden layers [18]. RBMs became foundational in deep learning, particularly in the unsupervised training of deep belief networks (DBNs), as demonstrated by Hinton, Osindero, and Teh in 2006 [19].

19.5.1 Mathematical Formulation of Boltzmann Machines

A Boltzmann machine is a type of stochastic recurrent neural network used in machine learning and statistical mechanics. It models complex data distributions by minimizing an energy function and learning patterns within data through unsupervised learning. This network is based on the principles of the Boltzmann distribution in statistical mechanics. A Boltzmann machine consists of binary nodes (neurons), each of which can be in one of the two states, usually represented as $s_i = \pm 1$ or $s_i = 0, 1$. The nodes are fully connected, with symmetric weights $w_{ij} = w_{ji}$, and include both visible nodes (input data) and hidden nodes (to capture latent features). The network learns a distribution over input data by minimizing its energy and converging to stable states corresponding to patterns in the data. The energy function for a Boltzmann machine given a state configuration $\mathbf{s} = (s_1, s_2, \ldots, s_N)$ is defined as

$$E(\mathbf{s}) = -\frac{1}{2} \sum_{i=1}^{N} \sum_{j=1}^{N} w_{ij} s_i s_j - \sum_{i=1}^{N} b_i s_i, \tag{19.12}$$

where w_{ij} is the symmetric weight between nodes i and j, s_i is the state of node i, b_i is the bias term for node i, and N is the total number of nodes.

The Boltzmann machine aims to minimize this energy function by finding configurations $\mathbf{s}$ corresponding to low-energy states, which represent patterns or correlations learned by the network. The probability of the network being in a particular state $\mathbf{s}$ follows the Boltzmann distribution:

$$P(\mathbf{s}) = \frac{e^{-E(\mathbf{s})}}{Z},$$

where $E(\mathbf{s})$ is the energy of the state $\mathbf{s}$ and Z is the partition function, defined as $Z = \sum_{\mathbf{s}} e^{-E(\mathbf{s})}$, which normalizes the probability distribution. This distribution ensures that lower energy states have higher probabilities, making them more likely to represent the patterns that the network learns.

19.5.2 Learning in Boltzmann Machines

Learning in Boltzmann machines involves adjusting weights w_{ij} and biases b_i to minimize the difference between the observed data distribution and the model's distribution. This is achieved through a gradient-based approach known as *contrastive divergence* and that proceeds in the following steps:

(a) In the *Positive Phase* the machine is clamped to an observed data sample, and correlations between node pairs are measured. (b) In the next *Negative Phase* the network runs freely, generating patterns, and correlations are measured again. (c) Finally, in the *Weight Update Phase* the weights are updated based on the difference in correlations between the positive and negative phases:

$$\Delta w_{ij} = \eta \left(\langle s_i s_j \rangle_{\text{data}} - \langle s_i s_j \rangle_{\text{model}} \right),$$

where η is the learning rate, and $\langle s_i s_j \rangle$ denotes the expected correlation between nodes i and j under the data (positive phase) and model (negative phase) distributions. This learning process allows the Boltzmann machine to adjust its parameters to model the observed data.

A *RBM* is a simplified form of the Boltzmann machine with connections only between visible and hidden layers, with no connections within the same layer. This structure significantly reduces computational complexity and is commonly used in applications requiring deep learning.

The Boltzmann machines have a number of very significant properties including energy-based modeling as in Hopfield networks. The energy function models the dependencies in the data, with lower energy states representing learned patterns. Furthermore, the nodes update probabilistically, allowing the network to model complex and multimodal distributions. Since they are able to learn patterns without labels, they represent an unsupervised learning method. Their weights are updated through the gradient of the likelihood function and usually approximated by contrastive divergence. Boltzmann machines, especially RBMs, have been fundamental for modern deep learning models, including deep belief networks (DBNs) and autoencoders, due to their ability to model complex, high-dimensional data distributions.

19.6 The Protein Folding Problem

The *protein folding problem* is one of the most significant unsolved questions in biochemistry and molecular biology. It asks how a protein's amino acid sequence determines its three dimensional structure, which is crucial for its biological function. Understanding protein folding has profound implications for fields like medicine, molecular biology, and biochemistry. The protein folding problem was

19.6 The Protein Folding Problem

first formally addressed in the 1950s and 1960s. Christian Anfinsen demonstrated that the amino acid sequence of a protein contains all the information necessary for it to fold into its native structure. This principle, known as Anfinsen's thermodynamic hypothesis, suggested that proteins fold spontaneously to minimize their free energy [20]. In 1969, Cyrus Levinthal pointed out that if a protein were to sample all possible conformations to find its native structure, it would take longer than the age of the universe to fold. This observation, known as the *Levinthal paradox*, highlighted the efficiency of the protein folding process, suggesting that folding follows a specific pathway rather than random sampling [21].

The development of *energy landscape theory* in the 1980s and 1990s provided a statistical physics perspective on protein folding. This theory describes protein folding as movement through a rugged energy landscape with multiple local minima. The concept of a "funnel-shaped" energy landscape helped explain how proteins achieve their native structure efficiently despite the vast number of possible conformations [22].

The protein folding problem remained unsolved for decades until the rise of computational methods and machine learning in the twenty-first century. In 2020, DeepMind's *AlphaFold* algorithm achieved remarkable success in predicting protein structures with near-experimental accuracy, a breakthrough that is widely seen as solving a large part of the protein folding problem. AlphaFold's success marked a significant milestone in computational biology and demonstrated the power of AI in solving complex scientific problems [23].

19.6.1 *AlphaFold: Solving the Protein Folding Problem with AI*

AlphaFold, developed by DeepMind, is an advanced artificial intelligence (AI) system that has significantly impacted structural biology by solving the protein folding problem, a major scientific challenge persisting for over 50 years. Protein folding is the process by which a protein's amino acid sequence adopts a specific three dimensional structure, essential for its function. Accurate prediction of protein structures traditionally required months of experimental work, involving techniques like X-ray crystallography and cryo-electron microscopy. In 2020, AlphaFold showcased unprecedented accuracy at the CASP (Critical Assessment of Structure Prediction) competition, achieving a mean Global Distance Test (GDT) score of approximately 92.4. This score reflects predictions with atomic-level precision, achieving median errors within 1 Angstrom (the width of an atom), which was previously considered computationally impractical.

AlphaFold achieved this milestone by utilizing a combination of deep learning and biological data, allowing it to generate highly accurate 3D structures of proteins in a fraction of the time required by experimental methods. This advancement opens doors for research in drug discovery, molecular biology, and the understanding of diseases. AlphaFold's methodology combines neural networks with evolutionary and structural biology data. It uses *multiple sequence alignment (MSA)* to identify evolutionary relationships and structural patterns within amino acid sequences. A *spatial graph* models physical relationships between residues, with nodes representing amino acids and edges representing their interactions. The model iteratively updates and refines the spatial relationships to predict the protein structure and assesses the reliability of each prediction.

In collaboration with the European Bioinformatics Institute (EMBL-EBI), DeepMind developed the AlphaFold Protein Structure Database.[1] This open-access database includes predictions for over 200 million proteins across various organisms, representing nearly the entire known protein universe. The database provides an invaluable resource with applications in areas such as (a) *Drug Discovery*: facilitating the identification of target sites for pharmaceutical interventions, (b) *Disease Research*: advancing studies in diseases such as malaria and neglected tropical diseases, and (c) *Fundamental Biology*: enhancing understanding of biological mechanisms through detailed protein structures. AlphaFold has already transformed structural biology by setting a new standard for computational accuracy and efficiency in protein modeling. Its success demonstrates the power of AI in scientific discovery, with implications beyond biology, including agriculture, environmental science, and health.

19.7 Conclusion

The coupling of statistical mechanics ideas with dynamics and computer science led to phenomenal advances over the last 50 years in science. The application of the newly discovered methods transcends specialized disciplines and embraces all sciences. The fruitful interaction among physics, chemistry, biology, and computer science led to a new synthesis that is founded on mathematics. The Ising model was initially introduced approximately 100 years ago for the study of the paramagnetic-ferromagnetic phase transition proved to be very fundamental also for the age of artificial intelligence [24]. The study of the random Ising model led to a better understanding of the behavior of interacting complex systems and enabled the development of mathematical techniques for their analysis. The Hopfield model that is founded on the random Ising model showed that with the inclusion of dynamics the resulting network can store efficiently memory patterns in nonlinear attractors. Boltzmann machines move the capabilities of these networks further and enable to not only amorphous memory storage but also labeling. And finally, with the advent of large and very sophisticated nonlinear networks as the AlphaFold, it is possible to address and solve outstanding, foundational problems in biology as the problem of protein folding. It is truly remarkable that in a seemingly unrelated path of studies the end result is such a powerful development in medical and biological sciences.

---

[1] https://alphafold.ebi.ac.uk/

19.8 Summary

- Spin glasses are complex statistical systems that lay the foundation for complex neural networks.
- The Hopfield model is a spin model that incorporates associative memory.
- Boltzmann machines are recurrent neural networks that learn through contrasting divergence.
- Restricted Boltzmann machines are simpler bipartite Boltzmann machines that are easier to train.
- The AlphaFold network of DeepMind was able to resolve the problem of protein folding and produce three dimensional structures of a large number of proteins.

References

1. The Nobel Prize, Press Release: The Nobel Prize in Physics 2021. https://www.nobelprize.org/prizes/physics/2021/press-release/
2. E. Ising, Contribution to the theory of ferromagnetism. Zeitschrift für Physik **31**, 253–258 (1925)
3. W. Lenz, Contributions to the theory of magnetic phenomena. Physikalische Zeitschrift **21**, 613 (1920)
4. S.G. Brush, History of the Lenz-Ising model. Rev. Modern Phys. **39**, 883–893 (1967)
5. L. Onsager, Crystal statistics. I. A two-dimensional model with an order-disorder transition. Phys. Rev. **65**, 117–149 (1944)
6. R.J. Baxter, *Exactly Solved Models in Statistical Mechanics* (Academic, New York City, 1982)
7. K.G. Wilson, Renormalization group and critical phenomena. I. Renormalization group and the kadanoff scaling picture. Phys. Rev. B **4**, 3174–3183 (1971)
8. S.F. Edwards, P.W. Anderson, Theory of spin glasses. J. Phys. F Metal Phys. **5**, 965–974 (1975)
9. D. Sherrington, S. Kirkpatrick, Solvable model of a spin-glass. Phys. Rev. Lett. **35**, 1792–1796 (1975)
10. G. Parisi, Infinite number of order parameters for spin-glasses. Phys. Rev. Lett. **43**, 1754–1756 (1979)
11. M. Mezard, A. Montanari, *Information, Physics, and Computation* (Oxford University Press, Oxford, 2009)
12. J.J. Hopfield, Neural networks and physical systems with emergent collective computational abilities. Proc. Natl. Acad. Sci. **79**, 2554–2558 (1982)
13. G.E. Hinton, T.J. Sejnowski, Optimal perceptual inference. In: *Proceedings of the IEEE Conference on Computer Vision and Pattern Recognition* (1983)
14. D.J. Amit, H. Gutfreund, H. Sompolinsky, Storing infinite numbers of patterns in a spin-glass model of neural networks. Phys. Rev. Lett. **55**, 1530–1533 (1985)
15. R. Schulz, R.C.J. Somers, *Associative Memory: Advances and Perspectives* (Academic, New York, 1994)
16. G.E. Hinton, T.J. Sejnowski, Learning and relearning in boltzmann machines, in *Parallel Distributed Processing: Explorations in the Microstructure of Cognition*, vol. 1 (MIT Press, Cambridge, 1986)
17. G.E. Hinton, Training products of experts by minimizing contrastive divergence. Neural Comput. **14**(8), 1771–1800 (2002)

18. P. Smolensky, Information processing in dynamical systems: Foundations of harmony theory, in *Parallel Distributed Processing: Explorations in the Microstructure of Cognition*, vol. 1 (MIT Press, Cambridge, 1986)
19. G.E. Hinton, S. Osindero, Y.-W. Teh, A fast learning algorithm for deep belief nets. Neural Comput. **18**(7), 1527–1554 (2006)
20. C.B. Anfinsen, Principles that govern the folding of protein chains. tScience **181**(4096), 223–230 (1973)
21. C. Levinthal, How to fold graciously, in *Mossbauer Spectroscopy in Biological Systems, Proceedings of a Meeting Held at Allerton House, Monticello, Illinois* (1969), pp. 22–24
22. J.D. Bryngelson, J.N. Onuchic, N.D. Socci, P.G. Wolynes, Funnels, pathways, and the energy landscape of protein folding: a synthesis. Proteins Struct. Funct. Bioinf. **21**(3), 167–195 (1995)
23. J. Jumper et al., Highly accurate protein structure prediction with alphafold. Nature **596**, 583–589 (2021)
24. A. Engel, C. Van den Broeck, *Statistical Mechanics of Learning* (Cambridge University Press, Cambridge, 2004)

Chapter 20
Computational Complexity and the Butterfly Effect

Are Neural Networks Stable Enough?

Abstract In this short chapter we summarize and give some remarks on the connection of AI with complex systems. While the universal approximation theorem guarantees using neural networks for approximating complex functions, the stability of the networks should also be taken into account.

20.1 Closing Remarks and Summary

In this book we embarked on a journey through two seemingly distant areas of knowledge: one is *Artificial Intelligence* and the second that of *Complex Systems*. The former developed, at least initially, from the interaction of computer science with neuroscience. The latter is a child of Newton seen first through the eyes of Poincare. The AI is now one of the most promising areas of science and technology with fantastic applications in engineering, medicine, social media, etc. Futuristic ideas about digital personas slowly take form and change our lives. Complex systems on the other hand developed over the last 50 years and the knowledge generated penetrated all science. Everybody knows about chaos and the fact that it is practically impossible to predict the future in certain important cases. Can we tell the weather in Boston one year from today? Not really, although we might get a reasonable idea statistically from previous knowledge. There is hope that by merging AI with complex systems we will both acquire better information on the latter and learn more about the former.

In this work we gave an exposition of both and attempted to show cases where the interaction with AI with complicated systems is quite beneficial. The presentation focused on a specific point of view and results primarily obtained from my research group in recent years. Clearly, there are many more examples and approaches that other groups and researchers have followed that are not included here. This book is not a review of the area; it has the main purpose to assist interested students and researchers to enter and work in the area. I strongly believe there is strong margin for future development and many great things will emerge in the not so distant future.

In the first parts of the book we gave an introduction to basic methods of machine learning. The main ideas were presented not only for machine learning without the use of neural networks but importantly with neural nets as well. Together with the techniques we also included a number of simple Jupyter Notebooks that may be used to work on the methods. The neural network part is quite important because the neural methods are very potent and can be used in many applications.

The second half of the book is rather specialized and presents the applications of the machine learning methods in complex systems and their applications. Special emphasis is paid in the nonlinear Schrödinger equation in its discrete form since analytical solutions are known for simple configurations. This is the test bed of the methodology, and since it works for analytically known cases, it is extended to unknown problems. The targeted energy transfer model is described in some detail and the methods used in this context.

The next sections focus on more complex situations that include dynamical chaos of few degrees of freedom and spatiotemporal chaos in extended systems. The use of autoencoders seems to be appropriate for learning features of chaos, while more complex methods augmented with partial use of ground-truth information are needed in chimeras and branching. An arsenal of methods including feed-forward neural networks, reservoir computing, and recurrent neural networks was seen to be very useful in predicting both chimeras and the onset of singular branching events. Localized nonlinear modes of the form of discrete breathers can also be recognized through the use of convolutional neural networks.

Moving into the quantum realm we showed that the quantum targeted energy transfer can be easily investigated through machine learning approaches. Furthermore truly quantum machine learning algorithms can be applied to qubit systems and assist in error correction.

In the domain of biomedical applications the connection of action potential propagation in the heart with the observed electrocardiogram provides a formidable task for both computational science and machine learning. We covered the model aspects of the problem and subsequently applied machine learning methodology in a case where we search for a connection between the electrical activity of the heart and hypertension. This machine learning analysis gives the surprising and useful result of their connection. Finally, we deal with real COVID-19 data, and through the use of physics informed methodology we show that machine learning can aid predictions in epidemiology.

Since 2024 was a phenomenal year where four scientists were awarded Nobel prizes for work in artificial intelligence, we could not but include a brief summary of the ideas and the research they did. It is noteworthy pointing out that this work is related to the Nobel physics prize awarded three years earlier for research in complex statistical and dynamical systems. The link between AI and complex systems is very real!

20.2 Digital Twins and Nonlinear Physics

A digital twin is a virtual representation of a physical system or process that mirrors its real-world counterpart. This digital model simplifies monitoring, simulation, and analysis. It also enables predictions and very importantly scenario testing without directly affecting the physical entity. Digital twins are widely used in industries like manufacturing, healthcare, aerospace, as well as in smart cities. They enhance efficiency and decision-making by providing insights and predictions based on the real time behavior and performance of the physical systems they replicate. Applying digital twins in nonlinear models replaces in some sense a complex system by an alternative perhaps equally complex. The latter may be more manageable however, and the use of the extended knowledge on the machine learning processes enables perhaps a more systematic observability.

20.3 Universal Approximation Theorem and Chaos

The Universal Approximation Theorem states that a neural network with a single hidden layer, containing a sufficient number of neurons, can approximate any continuous function on a closed and bounded subset of R^n, given appropriate activation functions (such as the sigmoid or ReLU) [1–3]. This theorem shows the theoretical power of neural networks, implying that they can model complex, nonlinear relationships between inputs and outputs. However, while the theorem guaranties the existence of such an approximation, it does not specify the number of neurons required or how to efficiently train the neural network. Therefore, practical considerations like network architecture, optimization, and training still play crucial roles in building effective neural networks. While the theorem warranties the approximation capabilities of complex neural networks does not give any input on how stable these networks may be. Given that the neural networks are complex systems themselves, it is very easy to imagine that in many cases they will have issues of stability that may affect their predictive power of the results [4]. As the networks become larger and larger, this problem might be more and more relevant. Research in this area will be very important and forthcoming.

20.4 Summary

- AI and Complex Systems can work together for mutual evolution.
- Digital twins are important both in practice and in theoretical models.
- Neural networks are complex systems themselves and can be analyzed using complex system's methods.

References

1. G. Cybenko, Approximation by superpositions of a sigmoidal function. Mathemat. Control Signals Syst. **2**(4), 303–314 (1989)
2. K. Hornik, M. Stinchcombe, H. White, Multilayer feedforward networks are universal approximators. Neural Netw. **2**(5), 359–366 (1989)
3. K. Funahashi, On the approximate realization of continuous mappings by neural networks. Neural Netw. **2**(3), 183–192 (1989)
4. M.J. Colbrook, V. Antum, A.C. Hansen, Can stable and accurate neural networks be computed? On the barriers of deep learning and Smale's 18th problem (2021). arXiv:2101.08286v2

Appendix A
Jacobi Elliptic Functions

Jacobi elliptic functions, denoted as $\text{sn}(u, k)$, $\text{cn}(u, k)$, and $\text{dn}(u, k)$, are important functions in the theory of elliptic integrals and arise in various areas of mathematical physics. These functions depend on two variables: the argument u and the elliptic modulus k, where $0 \leq k \leq 1$.

Let the incomplete elliptic integral of the first kind be defined as

$$u = F(\varphi, k) = \int_0^{\varphi} \frac{d\theta}{\sqrt{1 - k^2 \sin^2 \theta}},$$

where φ is the amplitude.

The Jacobi elliptic functions are then defined as follows:

- $\text{sn}(u, k)$: The sine amplitude function (Jacobi elliptic sine) is defined as

$$\text{sn}(u, k) = \sin(\varphi).$$

- $\text{cn}(u, k)$: The cosine amplitude function (Jacobi elliptic cosine) is defined as

$$\text{cn}(u, k) = \cos(\varphi).$$

- $\text{dn}(u, k)$: The delta amplitude function is defined as

$$\text{dn}(u, k) = \sqrt{1 - k^2 \sin^2(\varphi)}.$$

A Jacobi Elliptic Functions

A.1 Main Identities

A.1.1 Pythagorean-Like Identities

$$\text{sn}^2(u,k) + \text{cn}^2(u,k) = 1,$$
$$k^2 \text{sn}^2(u,k) + \text{dn}^2(u,k) = 1.$$

A.1.2 Derivative Relations

$$\frac{d}{du}\text{sn}(u,k) = \text{cn}(u,k)\,\text{dn}(u,k),$$
$$\frac{d}{du}\text{cn}(u,k) = -\text{sn}(u,k)\,\text{dn}(u,k),$$
$$\frac{d}{du}\text{dn}(u,k) = -k^2\,\text{sn}(u,k)\,\text{cn}(u,k).$$

A.1.3 Addition Formulas

$$\text{sn}(u+v,k) = \frac{\text{sn}(u,k)\text{cn}(v,k)\text{dn}(v,k) + \text{sn}(v,k)\text{cn}(u,k)\text{dn}(u,k)}{1 - k^2\,\text{sn}^2(u,k)\text{sn}^2(v,k)},$$

$$\text{cn}(u+v,k) = \frac{\text{cn}(u,k)\text{cn}(v,k) - \text{sn}(u,k)\text{sn}(v,k)\text{dn}(u,k)\text{dn}(v,k)}{1 - k^2\,\text{sn}^2(u,k)\text{sn}^2(v,k)},$$

$$\text{dn}(u+v,k) = \frac{\text{dn}(u,k)\text{dn}(v,k) - k^2\,\text{sn}(u,k)\text{sn}(v,k)\text{cn}(u,k)\text{cn}(v,k)}{1 - k^2\,\text{sn}^2(u,k)\text{sn}^2(v,k)}.$$

A.2 Special Cases

(a) When $k = 0$, the Jacobi elliptic functions reduce to elementary trigonometric functions:

$$\text{sn}(u,0) = \sin(u), \quad \text{cn}(u,0) = \cos(u), \quad \text{dn}(u,0) = 1.$$

(b) When $k = 1$, the Jacobi elliptic functions reduce to hyperbolic functions:

$$\text{sn}(u,1) = \tanh(u), \quad \text{cn}(u,1) = \text{sech}(u), \quad \text{dn}(u,1) = \text{sech}(u).$$

Appendix B
Weierstrass Elliptic Function

The Weierstrass elliptic function, denoted by $\wp(z; g_2, g_3)$, is a fundamental function in the theory of elliptic functions. It is a doubly periodic meromorphic function, meaning it has two independent periods, and is defined in terms of two constants g_2 and g_3, known as the Weierstrass invariants.

The Weierstrass elliptic function $\wp(z)$ is defined by the following series:

$$\wp(z; g_2, g_3) = \frac{1}{z^2} + \sum_{\omega \neq 0}\left(\frac{1}{(z-\omega)^2} - \frac{1}{\omega^2}\right),$$

where the summation runs over all nonzero lattice points $\omega = m\omega_1 + n\omega_2$, with $m, n \in \mathbb{Z}$ and ω_1, ω_2 being the fundamental periods of the lattice.

The Weierstrass elliptic function satisfies the following differential equation:

$$\left(\frac{d\wp}{dz}\right)^2 = 4\wp^3 - g_2\wp - g_3,$$

where g_2 and g_3 are constants related to the elliptic curve associated with $\wp(z)$.

B.1 Basic Identities

B.1.1 Derivatives

$$\wp'(z; g_2, g_3) = -2\sum_{\omega}\frac{1}{(z-\omega)^3},$$

where $\wp'(z)$ is the derivative of $\wp(z)$ with respect to z.

B.1.2 Differential Equation

$$\wp'(z)^2 = 4\wp(z)^3 - g_2\wp(z) - g_3.$$

This is known as the *Weierstrass differential equation*, and it characterizes the function in terms of g_2 and g_3.

B.1.3 Laurent Series

Around $z = 0$, $\wp(z)$ has the following expansion:

$$\wp(z) = \frac{1}{z^2} + \frac{g_2}{20}z^2 + \frac{g_3}{28}z^4 + O(z^6).$$

B.1.4 Addition Formula

The Weierstrass elliptic function satisfies the addition formula:

$$\wp(z+w) = -\wp(z) - \wp(w) + \frac{1}{4}\left(\frac{\wp'(z) - \wp'(w)}{\wp(z) - \wp(w)}\right)^2.$$

B.2 Special Cases

B.2.1 Near the Lattice Points

When z approaches a lattice point ω, the function $\wp(z)$ behaves like

$$\wp(z) \sim \frac{1}{(z-\omega)^2} \quad \text{as} \quad z \to \omega.$$

B.2.2 Special Values

At $z = 0$, the Weierstrass elliptic function satisfies

$$\wp(0) = \infty, \quad \wp'(0) = 0.$$

B.2.3 Elliptic Curve Relation

The Weierstrass elliptic function relates to elliptic curves of the form

$$y^2 = 4x^3 - g_2 x - g_3,$$

where $(x, y) = (\wp(z), \wp'(z))$ traces out points on the elliptic curve as z varies.

B.3 Connection to Other Elliptic Functions

The Weierstrass elliptic function can be expressed in terms of other elliptic functions. For example, it is related to the Jacobi elliptic functions via the following relations:

$$\wp(z) = e_1 + (e_2 - e_1)\operatorname{sn}^2\left(\sqrt{e_3 - e_1}\, z\right),$$

where e_1, e_2, e_3 are the roots of the cubic polynomial $4x^3 - g_2 x - g_3 = 0$.

Appendix C
Python Programming Language

Python is a versatile, high-level programming language that emphasizes code readability and simplicity. It supports multiple programming paradigms, including procedural, object-oriented, and functional programming. Python is widely used in web development, scientific computing, data analysis, artificial intelligence, machine learning, automation, and more. It has the following main features:

- *Easy to Learn*: Python has a simple, readable syntax, which makes it an excellent choice for beginners.
- *Dynamically Typed*: Python does not require variable declarations; data types are determined at runtime.
- *Interpreted Language*: Python code is executed line by line by the Python interpreter, making debugging easier.
- *Extensive Libraries*: Python has a rich ecosystem of libraries and frameworks, such as:
  - `NumPy` and `SciPy` for scientific computing
  - `pandas` for data manipulation and analysis
  - `matplotlib` and `seaborn` for data visualization
  - `Scikit-learn` for machine learning and data analysis
  - `TensorFlow` and `PyTorch` for machine learning and AI
- *Cross-Platform*: Python runs on various operating systems, including Windows, macOS, and Linux.
- *Large Community*: Python has a vast, active user base, making it easy to find resources, tutorials, and support.

C.1 Jupyter Notebooks

Jupyter Notebooks are interactive, web-based environments that allow users to create and share documents containing live code, equations, visualizations, and explanatory text. They are commonly used for data analysis, scientific research, and educational purposes. Basic features are:

- *Interactive Environment*: Jupyter Notebooks allow you to write and execute Python code in "cells" interactively. Each cell can be run independently, and the output is displayed directly below the cell.
- *Rich Text Support*: Notebooks support Markdown, allowing users to include formatted text, equations (using LaTeX), and images alongside the code.
- *Visualization Integration*: Libraries like `matplotlib`, `plotly`, and `seaborn` integrate seamlessly with Jupyter to produce inline visualizations.
- *Kernel Flexibility*: While commonly used with Python, Jupyter supports many other languages through kernels, including R, Julia, and Scala.
- *Reproducibility*: Notebooks can be saved and shared with others, preserving the code, output, and data analysis steps in a single file. This makes them ideal for sharing research and results.

C.1.1 Basic Workflow in a Jupyter Notebook

1. *Code Execution*: Write code in cells and execute them one at a time. For example:

```
x = 5
y = 10
print(x + y)
# Output: 15
```

C.2 Advantages of Jupyter Notebooks

- Interactive Learning: Perfect for teaching and learning, as it allows users to experiment with code and visualize results instantly.
- Data Science and Research: Commonly used in data science workflows, as it combines code, data, and visualizations in a coherent format.
- Documentation: Jupyter allows users to document their thought process, providing context and explanations alongside the code.

C.3 Installing Jupyter Notebooks

To install Jupyter, you can use the following command (assuming Python and `pip` are installed):

```
pip install Jupyter
```

To start a notebook, use the command:

```
Jupyter notebook
```

This will open the Jupyter Notebook interface in a web browser, where you can create and manage your notebooks.

Python, combined with Jupyter Notebooks, forms a powerful toolkit for interactive programming, data analysis, machine learning, and scientific research.

Appendix D
Introduction to TensorFlow

TensorFlow is an open-source machine learning framework developed by Google. It is widely used for building and deploying machine learning models, especially deep learning models. TensorFlow supports a wide range of neural network architectures, offers flexible deployment across multiple platforms, and is designed for both research and production environments. TensorFlow includes:

- *Scalability*: TensorFlow allows easy scaling from a single CPU or GPU to large distributed computing systems.
- *Automatic Differentiation*: TensorFlow computes gradients automatically, simplifying backpropagation for training neural networks.
- *Support for Various Platforms*: TensorFlow models can be deployed on cloud servers, edge devices, mobile devices, and even web browsers using TensorFlow.js.
- *Ecosystem*: TensorFlow includes libraries like TensorFlow Lite (for mobile), TensorFlow.js (for web), and TensorFlow Extended (for production workflows).
- *Tensor Manipulation*: TensorFlow efficiently handles multidimensional arrays (tensors) and provides a variety of operations for tensor manipulation.

D.1 Basic TensorFlow Workflow

TensorFlow models typically follow this basic structure:

- *Define the model*: Specify the architecture of the neural network, including layers, activation functions, and the number of units.
- *Compile the model*: Set the optimizer, loss function, and evaluation metrics.
- *Train the model*: Fit the model to training data by adjusting its weights to minimize the loss function.

- *Evaluate the model*: Use the trained model to make predictions, or evaluate its performance on test data.

D.2 Example: Linear Regression in TensorFlow

An example of the use of TensorFlow in linear regression:

```python
import tensorflow as tf

# Define the variables and placeholders
X = tf.constant([1.0, 2.0, 3.0, 4.0])
Y = tf.constant([0.0, -1.0, -2.0, -3.0])

# Define the model
model = tf.keras.Sequential([
    tf.keras.layers.Dense(units=1, input_shape=[1])
])

# Compile the model
model.compile(optimizer='sgd', loss='mean_squared_error')

# Train the model
model.fit(X, Y, epochs=500)

# Make predictions
print(model.predict([10.0]))
```

D.3 Introduction to Keras

Keras is an open-source, high-level neural networks API built on top of TensorFlow. It allows for easy and quick prototyping of deep learning models and provides a simple interface for building and training models with minimal code. The main features of Keras are:

- *User-Friendly API*: Keras is designed for ease of use, enabling fast experimentation with minimal code.
- *Modularity*: Models in Keras are built by stacking layers, allowing for flexibility and reuse of code.
- *Multiple Backend Support*: Although now integrated into TensorFlow, Keras was originally designed to run on multiple backends, including Theano and CNTK.
- *Extensibility*: Keras allows the creation of custom layers, loss functions, and metrics, making it highly extensible.

D.4 Basic Keras Workflow

A typical workflow in Keras consists of the following steps:

- *Define the model*: Create a sequential model and add layers, or use the functional API for more complex models.
- *Compile the model*: Choose an optimizer (e.g., Adam, SGD), a loss function (e.g., categorical cross-entropy), and metrics (e.g., accuracy).
- *Fit the model*: Train the model on the training data.
- *Evaluate the model*: Assess the model's performance on test data.
- *Predict*: Use the model to make predictions on new data.

D.4.1 Example: Building a Neural Network with Keras

This is how one can build a neural network with Keras:

```python
import tensorflow as tf
from tensorflow import keras

# Load the data (MNIST digits dataset)
mnist = keras.datasets.mnist
(train_images, train_labels), (test_images, test_labels) = mnist.load_data()

# Normalize the data
train_images = train_images / 255.0
test_images = test_images / 255.0

# Define the model
model = keras.Sequential([
    keras.layers.Flatten(input_shape=(28, 28)),
    keras.layers.Dense(128, activation='relu'),
    keras.layers.Dense(10, activation='softmax')
])

# Compile the model
model.compile(optimizer='adam',
              loss='sparse_categorical_crossentropy',
              metrics=['accuracy'])

# Train the model
model.fit(train_images, train_labels, epochs=5)

# Evaluate the model
test_loss, test_acc = model.evaluate(test_images, test_labels)
print(f"Test accuracy: {test_acc}")
```

```
# Make predictions
predictions = model.predict(test_images)
```

The Keras Sequential API is the simplest way to build models by stacking layers one after another. It is suitable for straightforward feed-forward models. The Keras Functional API is more flexible and can be used to create complex models, such as multi-input and multi-output models, directed acyclic graphs, or models with shared layers. TensorFlow, combined with Keras, forms a powerful framework for deep learning. TensorFlow provides low-level control and scalability, while Keras simplifies the process of building and training models.

Index

A
Action potential, 229
AlphaFold, 273
Artificial intelligence (AI), 3, 55, 256, 263, 277, 278
Artificial neural network (ANN), 55
Autoencoders, 83, 85, 145, 146, 216, 272, 278

B
Backpropagation, 55, 64
Boltzmann machine, 270
Branching, 177, 180

C
Cardiovascular, 235, 238
Chaotic systems, 6
Chimera, 164, 166, 171
Classification, 14, 27, 37
Complex systems, 4
Confusion matrix, 48
Conjugate layers, 212
Convolutional neural networks (CNNs), 56, 69, 188
COVID-19, 245, 247
Cross entropy, 59
Cross-validation, 51

D
Decision trees, 32
Deep learning, 64
Digital twin, 157, 279

Discrete breathers, 187, 189
Discrete Nonlinear Schrödinger (DNLS), 11, 97, 98
DNLS dimer, 121, 122
Dropout, 65

E
Early stopping, 66
Electrocardiogram (ECG), 223, 235, 236, 240

F
Feature selection, 239

G
Gini impurity, 32
Gradient descent, 44, 46
Graphene, 183

H
Hebb rule, 263, 269
Hopfield model, 268
Hopfield network, 263
Hyperparameters, 19, 64

I
Integrability, 5
Iris dataset, 29
Ising model, 263, 264

L
Learning rate, 44
Linear regression, 22, 45
Logistic regression, 28
Long Short-Term Memory (LSTM), 81
Loss function, 204
Lyapunov exponents, 146

M
Machine learning, 4, 12, 13, 21, 172, 194, 208, 235, 241
Mean absolute error (MAE), 50
Mean squared error (MSE), 50
Metamaterials, 159, 187

N
Neural networks, 211
Nondegenerate dimer, 122
Nondegenerate nonlinear dimer, 127
Nonlinear dimer, 135
Nonlinearity, 8

O
Observers, 173
Overfitting, 17, 25, 33, 51, 65, 218

P
Perceptron, 56
Physics informed machine learning, 84, 90, 250
PINN, 257
Principal component analysis (PCA), 39, 40, 42, 43
Protein folding, 272

Q
Quantum autoencoder, 212, 219
Quantum neural networks, 211, 215

Quantum targeted energy transfer (QTET), 199, 205
Qubit, 213, 217

R
Random forest, 237, 239
Rectified Linear Unit (ReLU), 60, 279
Recurrent neural networks (RNNs), 56, 69
Replica method, 267
Replica symmetry breaking, 266, 267
Reservoir computing, 83, 88
Restricted Boltzmann machine (RBM), 271, 272

S
Selftrapping, 98
Selftrapping transition, 122, 124
Sigmoid, 60, 79, 150, 279
Singular value decomposition (SVD), 42, 43
Soliton, 8, 97
Spatiotemporal, 278
Spin glass, 264, 266
SQUID metamaterials, 159, 163
Strange attractor, 145
Supervised learning, 14
Support vector machine (SVM), 33

T
Targeted energy transfer (TET), 132, 141, 199
TensorFlow, 202
Training, 58

U
Unsupervised learning, 15

V
Variational autoencoders, 87

The manufacturer's authorised representative in the EU is Springer Nature Customer Service Centre GmbH, Europaplatz 3, 69115 Heidelberg, Germany. If you have any concerns regarding our products, please contact ProductSafety@springernature.com

Printed and bound by CPI Group (UK) Ltd, Croydon, CR0 4YY

25/03/2026

02078171-0011